DATE DUE

MY 28'98			
AP 15 02			

DEMCO 38-296

MATHEMATICAL

CRANKS

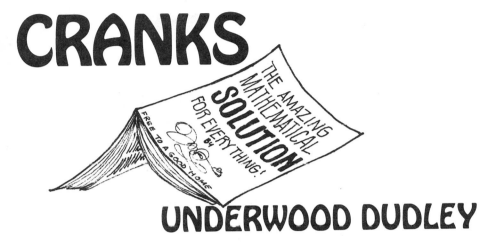

UNDERWOOD DUDLEY

THE MATHEMATICAL ASSOCIATION OF AMERICA
Washington, D.C. 1992

 MAA SPECTRUM

RIES

Published by
THE MATHEMATICAL ASSOCIATION OF AMERICA

———

©*1992 by*
The Mathematical Association of America (Incorporated)
Library of Congress Catalog Card Number 92-64179

ISBN 0-88385-507-0

Printed in the United States of America

Current printing (last digit):
10 9 8 7 6 5 4 3 2 1

INTRODUCTION

Well, what's the excuse for this *book?*

I guess that what's in it isn't in any other book.

Like what?

Things about mathematical cranks and eccentrics. No one pays much attention to them. Professional mathematicians are all too busy with more important business, and students of mathematics hardly know there are such people.

What exactly are you talking about?

Well, there are the people who think they have squared the circle—you know, constructed a square with the same area as a circle using only a straightedge and compass. That can't be done, but circle-squarers think they've done it. Then there are the angle trisectors, who think they can divide angles into three equal parts with straightedge and compass, also impossible; and the people who think that they've proved Fermat's Last Theorem, that $x^n + y^n = z^n$ doesn't have any positive integer solutions when n is bigger than 2. That's not necessarily impossible, but they're all wrong too.

So this is a book about nuts?

No, no. They're not nuts. Well, a few are, but most aren't. A lot of them are amateurs—mathematical amateurs who don't know much mathematics but like to work on mathematical problems. Sometimes, when you can't convince them that they haven't done what they thought they've done, they turn into *cranks*; but cranks aren't nuts, they're just people who have a blind spot in one direction. Also, there's a lot in the book that isn't about nuts or cranks. There are the people who say that ten isn't the best base for the number system—they're right, but since we will never change bases, to bother about it is eccentric. There's the priest who prays in matrices, which is a little unusual, but not cranky. You can pass these people in the street and never know it.

All right. It's about nuts, cranks, and eccentrics. Does that describe what's in it?

And what they've done. There's a huge variety. There's how the college president who proved Euclid's fifth postulate managed to prove it, how someone found the circumference of an ellipse without elliptic integrals, and how someone else showed how to color maps with only *two* colors. There's also background material. If you want to trisect an angle or duplicate the cube using a straightedge with two marks on it, that's there; and there's a little history of squaring the circle and how to do it with the quadratrix of Hippias, things like that.

I see. So why would anyone want to read it?

Well, cranks are *interesting*. You never know what you're going to run into. There are a lot of surprises, and a lot of new ideas. Not all of the ideas are good, but they're new. Also, it can be useful to read about cranks in case you ever run into one: you'll know how he's likely to act, and you'll have seen enough examples of cranks to know what you should do. Even if you never meet one, you can see what has happened between cranks and other people. You can get to see how cranks' minds work. It's like going to the movies: you get to experience a whole new part of life with hardly any effort and no danger. You may even be able to understand cranks. I've been trying for twenty-five years or so and haven't made much progress, but someone else may do better. Cranks can be fascinating.

What would you say if I said that cranks are odd and anyone who's interested in them is probably a little peculiar too?

What? But....

Never mind. Where'd....

Wait a minute. Kenneth May, the historian of mathematics, once told me that he thought crank mathematics was the folk mathematics of its time and worth looking at. Why, even a crank, one of the people who proved Fermat's Last Theorem (he thought), wrote me: "The study of this material is a legitimate field of research. What is their basic psychology? Is it some form of ego defense? What types of error are they subject to?" It's no more peculiar than a lot of things.

The question was hypothetical. Where'd you get all this stuff?

It's a miscellaneous collection, since it's impossible to be systematic. Years ago I made an effort to gather data—writing to 600 departments of mathematics, going around to libraries—and that turned up some things (but not all that many, since most crank material gets thrown away), but most of it came my way from various sources. I'd like to thank some of the people who have, at one time or another, sent me material or information: Gerald Alexanderson, George E. Andrews, P. T. Bateman, Michael Bleicher, R. P. Boas, Andrew Bremner, James O. Brooks, Paul Campbell, Roger Cooke, Larry Curnutt, Gerald Folland, Martin Gardner, Robert E. Greenwood, Branko Grünbaum, Eugene Jacobs, Peter D. Johnson, Jr., Stephen King, Loren Larson, George E. Martin, Leroy Meyers, Zane Motteler, C. S. Ogilvy, Haim Reingold, Kenneth Ross, Doris Schattschneider, Martha Siegel, David Singmaster, J. A. Spencer, Michael Steuben, Ian Stewart, John Tinker, John R. Tucker, Harry Waldman, Gordon Walker, Alfred Willcox, Gerhard Wollan, Fritz Wolf, and Leon Zukowski.

I'd also like to thank the Fisher Faculty Development Fund of DePauw University for providing a semester free of teaching duties.

Yes, yes, let's get on with it. How much mathematics do you have to know to be able to read the book?

The readers I had in mind were people who had studied some mathematics at some time, and liked it. They don't have to be mathematics majors or anything like that, or have the derivative of the secant at their fingertips, but of course the more mathematics they know the better. Sometimes mathematical things are explained, but other times they are not, since the explanations would take up too much space and the purpose of the book is not to teach mathematics. If something in a section is unfamiliar or if the section is dull, then skip it—what comes next may be better.

"Section": how's the book organized?

Hardly at all. Since it's a miscellany, there are fifty or so sections, some long and some short, on one topic or another. I couldn't think of any better way to arrange them than alphabetically by topic.

By the way, some of the sections contain opinions of mine. They are not all necessarily shared by the publisher, the Mathematical Association of America, or by its membership. But people and opinions are interesting, don't you think?

I'll ask the questions. So this isn't a work of scholarship?

Not really. In a sense it is since it gathers together and comments on the writings of a good many authors, but its purpose is to be informative and interesting, not scholarly. There aren't any footnotes or bibliographies, or anything like that. There are some notes at the end that refer to places where anyone who is interested can look for more information on some of the things that are mentioned in the text. Those aren't systematic either.

Also, many cranks never publish their material for distribution to the general public. That is why I refer to such cranks by their initials only, and disguise or do not mention where they live. They don't appear in the notes. I use initials even for cranks who have died, and for those whose works have been published. It's really not important who they are. But the notes will usually include details on published books or pamphlets, especially when they were put out by commercial publishers.

That's considerate of you. Do you have anything else to say in favor of the book?

Well, no, I guess. . . . I can't think of anything.

Really? Nothing else?

No.

I see. Thank you, you may go.

CONTENTS

ALPHABET, APPLICATION OF PYRAMID HEIGHT TO THE

Many cranks give mathematics a great deal more credit than it deserves. One crank thought that the equation he discovered established beyond question the triune nature of God, another thought that his work

> should be studied for its applications to faster-than-light rocket travel as well as possible applications to teleportation and telekinesis

and a third thought that his trisection of the angle would enable gold to be made out of lead. Mathematics is powerful, but not *that* powerful.

Less extreme are those who claim only that numbers give information about non-numerical things, stopping short of asserting that they can work miracles. Among these are the numerologists, and among the numerologists are the pyramidologists. Pyramidologists are convinced that measurements of the Great Pyramid in Egypt contain an immense amount of valuable information. They think that the pyramid builders, in possession of ancient and occult wisdom now lost, constructed their pyramids so that they incorporated, in coded form, prophecies of events to come; and that, by deeply studying their structures, pyramidologists will be able to decode them and thus know the future.

Pyramidology had its heyday about sixty years ago and it has subsided since, being replaced by other methods of learning about the unknown such as UFOs and channeling. Fads come and go—the lost continent of Atlantis is not in the news as much as it used to be, and the number of advertisements placed in magazines by the Rosicrucians (A.M.O.R.C.) seems to have declined—but old ones linger on and there are at least a few pyramidologists still upholding the faith.

O. P. was a pyramidologist who showed in 1943 that the builders of the Great Pyramid knew that in English words the letter q would always be followed by the

1

letter *u*. P., who lived on North Euclid Avenue in Pasadena, demonstrated this in *Pyramid Height Fits the English Alphabet*, where he first showed that the pyramid builders knew there would be twenty-six letters in the English alphabet:

> The *King's Chamber* goes with the *United States*, hence it is not surprising that the *Pyramid height* is in harmony with the *English Alphabet*. The UP-joint numbers in the Ascending Passage have "28" as the last value, but there are *two VACANT* joints, thus showing "26"—"26" letters—regular joints. In the first chapter of S. Matthew there are *25 verses* with "Chapter 1" checking in the total as "*Z*".

The height of the pyramid, P. said, is 5813 pyramid inches. P. then numbered from 1 to 25 the letters of the alphabet from A to Y, leaving out Z for the reason pointed out above. Next, P. constructed a number pyramid with one row for each letter:

A							A						
B						1	2	1					
C					1	2	3	2	1				
D				1	2	3	4	3	2	1			
E			1	2	3	4	5	4	3	2	1		
F		1	2	3	4	5	6	5	4	3	2	1	
G	1	2	3	4	5	6	7	6	5	4	3	2	1

$$\cdots$$

continuing on to the row for Z, 1 2 . . . 24 25 24 . . . 2 1. The first row does not contain a number because, P. said, A is the pyramid letter—just look at its shape. To get the height of the pyramid, add together the heights of the rows. The heights of the rows are the sums of the numbers in them: 4, 9, 16, 25, . . . , 625, which are the squares of the integers from 2 to 25. P. then summed the squares by putting them in a column and adding, something we can do more quickly because we know a formula:

$$2^2 + 3^2 + \cdots + 25^2 = \frac{25 \cdot 51 \cdot 26}{6} - 1^2 = 5524.$$

Then P. related that number to the pyramid height. The 17th letter of the alphabet is *q* and the 21st is *u*. Their rows in the pyramid sum, respectively, to 289 and 441. The pyramid builders knew that *u* would always go with *q*, so we need to replace *u*'s 441 with *qu*'s 289 + 441 = 730. When we do that, the pyramid sum becomes

$$5524 + 289 = 5812,$$

the height of the actual pyramid. That can hardly be a coincidence and so must have been planned by the pyramid builders. As confirmation, P. noted that 730 is twice the number of days in the year. By the way, one of the reasons that the pyramid height is 5813 pyramid inches is that 5, 8, and 13 are Fibonacci numbers. Another is that $\frac{5813}{10\phi}$, where $\phi = \frac{1+\sqrt{5}}{2}$ is the Golden Ratio, equals 359.26, almost the number of degrees in a circle.

P. was quite serious, and his reasoning is typical of that employed by numerologists. In the minds of numerologists, numbers have magical powers that influence the non-numerical world. P. may have thought that it was *because* of the pyramid that English words always have u's following q's.

P. included a footnote:

Reason for building the Pyramid, to show the coming Star-Shift: Wednesday, Nov. 17, 1948.

The pyramid let him down there.

AMERICAN REVOLUTION, THE ROLE OF 57 IN THE

A small 24-page pamphlet, *History Computed*, with subtitle *America's 57 Constant*, was published in 1983 by A. F. It begins

> History and arithmetic are not expected to coincide and there is nothing on record showing unity between the two. This present success is therefore something new under the sun. Begun with mild curiosity the search for numerical coincidence was rewarded beyond any conceivable expectation.

F. had noted that four of the first six presidents of the United States—Washington, Jefferson, Madison, and John Quincy Adams—were inaugurated at age 57 (Monroe, president number five, just missed at age 58) and that *none* of the succeeding presidents was 57 years old when inaugurated. Something was going on there:

> That number, singly or in multiples, consistently marked events including the Declaration of Independence itself. Against the laws of probability, history and arithmetic came together making possible precise computations which until now were hidden in the record.

Here we go:

> The Boston Tea Party, Dec. 16, 1773, established the 57 constant. According to eyewitness accounts and today's plaque at the wharf site, 342 (6 × 57) chests of tea were destroyed. Events were to be placed by

that number. An organizer of the raid, William Molineux, 57, was its oldest member. The harbor water was covered with tea leaves from the 342 chests, a symbolic mantle for the 342 (6 × 57) men who were to die in naval action of the Revolution. The Department of Defense casualty figure is accompanied by 114 (2 × 57) wounded.

The steps leading from the Tea Party to the Revolution proceeded within a timetable of 57's.

On Feb. 11, 1774, 57 days after the Tea Party, its participants were charged with treason.

The text of Parliament's March 14, 1774, reprisal to the Tea Party was published in Boston 57 days later on May 10.

On May 12 the Boston Committee on Correspondence rejected Britain's demand for payment for the 342 chests of tea exactly 342 (6 × 57) days before April 19, 1775, when the Minute Men were to face British troops at Lexington and Concord.

On Sept. 5, 1774, the First Continental Congress was opened, 285 (5 × 57) days before the Battle of Bunker Hill, June 17, 1775.

You may be thinking there is nothing but coincidence in all those 57s and that if you had a good chronology of the Revolution you could produce as many similar facts for any other small integer. But there is more:

The sequence of the alphabet also was in harmony with Coincidence 57. A computer program to translate letters of words into alphabetical sequence numbers (A as 1, B as 2, etc.) will show that the prominent names in the Revolution, individually or in combination, invariably produce numbers totalling 57 or multiples of that number. Typing out on the computer keyboard the words Boston Tea Party-Revolution translates into figures totalling 342, the six times 57 tea chest number. Minute Men is 114, double 57. From that introduction the letters-numbers theme recurred through the background of the war.

The process of turning names and words into numbers is known as *gematria*, and has a long history. The ancient Greeks did not have the idea of inventing separate symbols for numerals—they used letters instead—so in Greek all words had numerical values. That is not true in English, but computers make it easy to calculate the number of a word and practice modern gematria. Even without a computer it is possible. For example, turning letters into numbers as F. did by replacing A with 1, B with 2, and so on to Z with 26, F. could have noticed that history itself proclaimed his 57:

H	8
I	9
S	19
T	20
O	15
R	18
Y	25

$$114 = 57 + 57.$$

F. went on:

Fifty seven days after Lexington and Concord, Congress on June 15, 1775, chose George Washington to command the new Continental Army. The choice came exactly 57 weeks after the May 12, 1774 defiance of the British ultimatum.

His great grandfather, John Washington, came from America to England in 1657 in the 57th year of the East India Company (chartered Dec. 31, 1600) whose tea was to be destroyed. Arrival of the first Washington was 228 (4×57) years before 1885 when the Washington Monument was to be dedicated.

Fifty seven years after 1657 another great grandfather also in a strange land arrived in England from Germany in 1714 to be crowned George I. His great grandson, George III, was to reign during the American Revolution. The alphabetical sequence numbers in the name England total 57. The two great grandsons were to be major adversaries in the Revolution and to have in common the 57 count in the name George.

Mary Washington, whose given name also totals 57, gave birth to George in 1732, 57 years before he was to become president in 1789 on April 30, 57 days after the first U. S. Congress was called on March 4. She was to live another 57 years after his birth.

A 1759 wedding announcement carried the names George Washington Esq. and Martha Dandridge Custis. Each name has an alphabetical sequence count of 228 (4×57). For part of the very first presidential year the ages of both the First President and the First Lady were 57.

There is more in the 24 pages of F.'s pamphlet. The letters in *United States of America* sum to four 57s when translated into numbers. The Marquis de Lafayette, born in 1757, left France to join the Revolution 57 years before his death, and the *Star-Spangled Banner* was written when he was 57 years old. The Revolution lasted for 3192 (fifty-six 57s) days. The version of the Declaration of Independence approved by Congress contained 1332 words.

Lulled by the seemingly endless succession of 57s, you might not notice that 1332 divided by 57 leaves a remainder of 21 and thus does not seem to fit into the pattern. If you think that, it is because you have not seen as deeply as F. did. He proceeded to show where the 1332 came from: start with 3192, the number of days in the Revolution, and note that

$$123 + 234 + 345 + 456 + 567 + 678 + 789 = 3192.$$

That is startling enough, but note in addition that 456 is 8 times 57, and also twice the numerical value of the letters in *United States of America*. Moreover,

456, derived from the double United States of America, is the keystone of an arch constructed of the seven numbers in the equation that measured the days of the Revolution.

$$
\begin{array}{ccc}
 & 456 & \\
345 & & 567 \\
234 & & 678 \\
123 & & 789
\end{array}
$$

Through the triumphal arch marches a column composed of the differences between the two sides:

$$
\begin{array}{c}
222 \\
444 \\
666
\end{array}
$$

The orderly array totals 1,332, the number of words in the title and text of the Declaration of Independence.

There you have the 1332. And if that is not enough,

The coincidence number digits 5 and 7 total 12 which, subtracted from each of the seven equation numbers, leaves 111, 222, 333, 444, 555, 666, and 777. While all the even triplets combine to the 1,332 word count of the Declaration, it happened also that all the odd triplets total 1776, the independence year.

Surely there must be *something* behind all that. However, F. drew no conclusions. The last paragraph of his pamphlet states:

This version of history began with the four 57's listed in a column on a chart. They total 228 and the four presidents' order of service numbers in the first column, 1, 3, 4 and 6, total 14. Fourteen times 228 equals 3,192 (56 × 57), the days of the Revolution. 228 divides evenly into the hours, minutes, and seconds from Lexington to Yorktown besides coinciding with other aspects of American history, including Jefferson's UNITED STATES OF AMERICA.

That F. did not go on to construct a theory of history based on 57, perhaps involving astral beings vibrating to 57 and influencing humans to vibrate that way also—why else did Mr. Heinz have 57 varieties?—with warnings to watch out for presidential candidates who would be 57 years old at inauguration and to be especially alert in 1995 (57 × 35; in 1938, 57 × 34, World War II was about to break out in Europe), is a tribute to his ability to resist the seductive power of mysticism. He merely presented his findings, letting readers see in them what they will.

What I see in them is a splendid refutation of pyramidologists, Stonehenge buffs, and others who deduce astonishing things from measurements of ancient monuments. What F. has demonstrated, and demonstrated very well, is that given enough numbers—whether measurements of pyramids or dates in history—it is possible to squeeze out of them things that may seem remarkable to people who do not realize that, given enough numbers, you can squeeze almost anything out of them.

Of course, if the United States does come to an end in 1995, we will have to admit that F. was really onto something and that we should have paid more attention. By the way, if the end of everything does not come in 1995, look out for 2004, because 2004 is 1776 plus four 57s.

APPLIED MATHEMATICS

Mathematics has applications. Science needs and uses mathematics, technology needs and uses science, and we all need and use technology. The process of applying mathematics is to construct a mathematical model of some segment of the world, apply mathematics to it to get results, and take the results back to the world to see if they work. Sometimes they do and sometimes they don't—models designed to predict the future of the economy seem to fail as often as they succeed—but the number of successes is large enough to keep people working on models.

Most mathematical model-makers realize the limitations of their craft, but cranks do not. One of the things that makes a crank a crank is his refusal to see what is perfectly obvious to everyone else. Mathematical models do well in physics and chemistry, but less well in economics and politics. When models must take the behavior of people into account, precision goes away. The reason for this is that the mathematics used to make models is so simple that it cannot deal with anything at all complicated. Algebra, calculus, and probability are really *simple* subjects, so of course they cannot be applied to complicated matters. Cranks do not realize this and thus feel free to apply mathematics not only to economics but to philosophy, religion, anything.

L. B., the author of

The Fundamental Equation of the Construction of the World

(published by the author in 1939), applied mathematics to *everything*. The subtitle of his pamphlet is

Or, The Analysis of the Human Mind

with an explanation

9

A Treatise on the Mathematical Solution of All Economic, Political, Religious, and Scientific Problems

When he said all, he meant *all*:

> It is surprising that the most simple phenomena of nature have not been subjected long ago to a scrutinizing mathematical investigation. Such an investigation is carried on in the following and it is fully justified to give it the bold title: The Fundamental Equation of the Construction of the World. With this equation all sciences including religion, psychology, and particularly economics will be based on a mathematical foundation. The reason is obvious, all those sciences involve the human mind, and our fundamental equation is nothing else but a representation of the functions of the human mind and their combination with nature.

Of course, there will have to be a revolution in human thinking:

> The great mathematician Pythagoras, several hundred years before Christ, claimed that the world with all its laws and phenomena could be described by figures or numbers. This has been accomplished in the fields of physical and technical sciences. All units in physics can be shown to be functions of the centimeter, the gram, and the second. The Centimeter-Gram-Second System however fails when basic problems of Economics, Religion, Philosophy, or Administrative Problems of Government are involved. An equation which claims to be the Fundamental Equation of the Construction of the World must solve all these problems, otherwise it has no right to such a claim.
>
> In order to solve all these problems it is necessary to establish a new system, the Position-Operation-Time-System, or the P-O-T System. This System is superior to the Centimeter-Gram-Second System or the C-G-S System. It is even questionable whether the C-G-S System will now collapse or will have to make place to a substantially modified new C-G-S System. It is even doubtful that any further basic new inventions will be made until this modification has been performed, since the present scientific research seems to be cornered as far as the angle of speculative thinking is concerned.

Let us have no obvious witticisms about P-O-T. The P-O-T system can be applied to philosophy:

> The Position-Operation-Time-System is the fundamental system of philosophy, and all definitions of philosophy must be represented as functions of position, operation, and time, in the same manner as all electrical, mechanical, or thermodynamic terms are given as functions in the

centimeter-gram-second-system. A philosopher who proceeds otherwise will find himself turning around in circles, in other words his philosophy will make no sense. . . . The relation between the position-operation-time functions is the fundamental equation, which is the basic equation of philosophy. The philosophy derived from the Fundamental Equation resembles the philosophies of Plato, Aristotle, and St. Thomas Aquinas.

To morality and religion:

It has been tried by governments and dictators to stamp out religion. This is an impossible and absurd undertaking. The reason why can be found in the fundamental equation. By means of this equation a part of religion will be turned into a mathematical science. The morality of an action can be defined by means of the fundamental equation. Coincidence with the fundamental equation is morality and deviation is immorality. The equation may be termed the basic equation of justice. With the aid of a newly discovered law, the law of maximum simplicity, which pervades the whole of nature, it is possible to derive the Ten Commandments from the fundamental equation. The highest perfection of the fundamental equation is an expression of the Trinity Nature of God.

To psychology:

Human thought as well as human thinking is embedded in matter. It can be reduced down to a counting process. The basic law by which this matter is assembled and disassembled is the fundamental equation. With a proper representation of the fundamental equation, which represents this counting process it will be possible to disclose a great variety of new laws, and psychology also will be a mathematical science.

To medicine:

The human body is made up of molecules, atoms, ions, and electrons. A particular cooperation of all those elements constitutes health. A deviation from this cooperation represents sickness. The basic law governing the behavior of all these elements is the fundamental equation.

To economics:

The wildest mix-up can be found in the teachings of political economy. This is a lamentable fact, since the happiness of millions of people depends on the proper application of the teachings of economics. The fundamental equation permits for the first time to give a correct definition of wealth, capital, interest, income, etc. The fundamental equation itself gives a definition of wealth, while the definition of capital can be found as a function of position, operation, and time of the fundamental equation. The same holds for all other terms of economics. All laws of economics can be derived from the fundamental equation in a similar procedure as the laws of mechanics and the motion of celestial bodies can be found from the law of gravitation.

To government:

> Justice and freedom of the people are the basic points of the Constitution of the United States. In this the fundamental equation is identical with the Constitution. The field of government in which justice and freedom should be manifested is the field of taxation. Government can be divided into two fundamental functions. The first one is the collection of funds or taxation, and the other one is the spending of funds or administration. If there should be any justice or freedom, both of these functions must be governed by the fundamental equation. . . . It is the fundamental equation which will balance the budget of the government within a month when properly applied.

That is *some* equation.

B. was not a lunatic, producing his ravings in an asylum. He billed himself as "Member of the American Society of Mechanical Engineers" and was prosperous enough to bear the cost of publishing his works himself. He was also the author of a long book, which he advertised in his pamphlet with

Important Announcement!

The most outstanding book of the 20th century will be ready for distribution in a few weeks. Secure your copy now.

THE MATHEMATICS OF UNLIMITED PROSPERITY

THE TRIUMPH OF MAN OVER NATURE

by the same Author

The reader of average intelligence will find this book extremely fascinating. All mathematical formulae are condensed in the last part.

The faults and shortcomings of the present economic system are developed as deviations from the fundamental equation.

For the first time proper definitions are given for: capital, interest, income, etc.

The unity of the whole world is discussed in the philosophical part.

The harmony between religion, science, and economics is derived from the fundamental equation.

The Trinity of God is derived from nature.

This book is predominately ECONOMIC and contains or indicates the mathematical solution of all economic problems.

406 pages with 23 diagrams, nomographs, and tables.

How did B. appear, I wonder, to those who associated with him every day and did not know about his discovery? Was his grandiosity all concealed under a plain exterior or did his inward fire show through? How common is his type?

By now, you are probably impatient to see the fundamental equation of the construction of the world, so useful in so many fields. Before disclosing it, B. gave several examples to help in its interpretation. Here is one, to be kept in mind:

THE EXAMPLE OF THE WAITRESS

A steady customer, representing the laws of nature, walks into a restaurant. An attractive young waitress, representing the human mind, serves a plate of soup as ordered by the customer. That is interest of position, the first action of the waitress. The customer, the laws of nature, eats the soup with delight. Most of the soup travels into his stomach, but some of the soup splashes over the rim of the plate. It can also be observed that a very small fraction of the soup volatilizes. The customer has not finished yet, and the attentive waitress serves another ladle of soup with a smile. This is interest of operation, the second action.

Note that the waitress never removes soup from the plate. Her whole activity consists in filling, that is the first filling and all the following fillings. Note also that the customer never fills, but always takes out. The laws of nature always derange that construction which had been established by the mind. It requires continuous activity of the mind to sustain the construction which has been established by the mind. If seven elements of construction, say seven atoms, have been arranged in a particular position, say in a straight line, it requires continuous activity of the mind to keep them there, in the same way as it takes continuous activity to keep the container filled to a certain mark.

Now to the fundamental equation. If E is an item, x the outside cause, and y the inside cause, then the total change in E is given by

$$dE = dE_x + dE_y,$$

Although fundamental, that is not yet the fundamental equation.

Both of the above functions, outside cause x and inside cause y, naturally are a function of the time t. The outside cause is a function of what is in existence on the outside multiplied by a factor p. The factor p is the rate of interest or operation. . . . The size of the change from the inside must be in proportion to the total in existence, times the factor q. This factor q is the rate of outflow or decomposition, and as such it is always negative. . . . Thus we obtain the equation

$$dE = E_x p\, dt + Eq\, dt \qquad (2)$$

This is the fundamental part of the fundamental equation. If you agree that

$$dE_x = E_x p\, dt \quad \text{and} \quad dE_y = E\, q\, dt,$$

then all the rest follows. If you understand those equations properly, then the rest is plain sailing. Divide both sides of (2) by dt and differentiate to get

$$\frac{d^2 E}{dt^2} = \frac{dE_x}{dt} p + \frac{dE}{dt} q.$$

Then eliminate E_x, using

$$\frac{dE_x}{dt} = \frac{dE}{dt} - \frac{dE_y}{dt} = \frac{dE}{dt} - E\,q$$

to get

$$\frac{d^2 E}{dt^2} - (p+q)\frac{dE}{dt} + p\,q\,E = 0.$$

This pleasingly symmetric differential equation can be solved by any student sufficiently far along in a course in differential equations to give

$$E = C_1 e^{pt} + C_2 e^{qt},$$

where

> The coefficient C_1 is the position of the mind, and the coefficient C_2 is the position of nature. Nature has no position of its own; it was impressed on it by the mind. The values of C_1 and C_2 are given by the conditions at the beginning and end of the first unit of time. Thus we obtain

$$E = \frac{E_1 - E_0 e^q}{e^p - e^q} e^{pt} - \frac{E_1 - E_0 e^p}{e^p - e^q} e^{qt}$$

wherein E_0 is the condition at the beginning of the first element of time
and E_1 that at the end.

That is it, the fundamental equation of the construction of the world.

And that is where B. stopped. He added a bit about how p and q do not have
to be constant with time and about how the fundamental equation would look when
p and q are variables, but he did not give any examples of its amazing applications.
The derivation of the triune nature of God would be especially interesting, since it
implies that Christianity is the only true religion, but there are no details. Nor are
there details of any of the other applications. B. probably meant his pamphlet mainly
as a teaser for his book ("price $4," rather a lot for 1939) and that is why he left them
out. And the reason he did not give a simple derivation of one of its consequences—
the Eighth Commandment, perhaps—is that he might have realized, deep down,
that it would not be convincing and would not stimulate sales of his book. Lacking
a copy of it, we will never know how the fundamental equation demonstrates that
laissez-faire economics is built into the world:

> It is above all necessary that the government keep out of business
> and abstain from all types of communistic management. Nobody else
> is in a better position to manage his own business than the individual
> himself. It requires also that the government remove all unjust taxation on
> business as well as income. The removal of interference and the removal
> of taxation are basic requirements of personal liberty to the benefit of the
> manufacturer as well as the working man.

Nor will we know how the equation

> permits the analyzation and solution of all problems in which the human
> mind is involved. This includes all laws and definitions of philosophy, all
> terms of natural religion and morality, and problems of psychology.

What a loss!

Of course, it is not a loss. No equation can do such things, especially one that is
a solution of a linear second-order differential equation with constant coefficients,
and only a crank would think that it could. Second-order differential equations can
describe simple things such as how electrons circulate in circuits with a few resistors
and capacitors scattered around in them, but the human mind and the problems it
causes are considerably more complicated. Economics cannot be reduced to an
equation, nor can religion.

Where did B. get the idea that an equation could have such power over nature?
Where do people get the idea that mathematics has such potency? It cannot be from
the applications of mathematics that they see in school. From solving problems
about how to mix peanuts at $2 a pound and cashews at $5 a pound to produce a

mixture to sell at $3 a pound, to solving problems about the ultimate purpose of human existence, is too large a leap to come naturally. Also, mathematics teachers are too honest to represent their subject as being able to do any more than it does. People who come to the erroneous conclusion that mathematics has such tremendous applications must do it by themselves, each person independently of all the rest. I suppose they do it because they know that the mathematics they saw in school solved problems in some mysterious way that they did not understand; so naturally advanced mathematics, even more mysterious, must be able to solve advanced problems. You would think that a mechanical engineer, able to solve linear second-order differential equations with constant coefficients, would know better, but he did not: evidently the prestige, authority, and illustriousness of mathematics is so great that even B. was dazzled unto blindness.

A. W. was another person who thought that his work could be useful. He sent it out as separate sheets, each one self-contained. One of them is headed

WORLD'S MOST COMPLETE

MATHEMATICS LIBRARY

is the

KNOWLEDGE

for

use

in

ALL INDUSTRIES

It goes on:

ENOUGH MATHEMATICS TO DO

ANYTHING THAT CAN BE DONE

and concludes:

> The nation, that makes total use of this COMPLETE MATHEMATICS KNOWLEDGE, can stay ALWAYS AHEAD of all military rivals, and ALWAYS AHEAD of all competitors for the world markets.

Most of his sheets contain material similar to the following, copied from one headed "1 RELATIONS"

> MATTER changes continuously, from CUBES to SPHERES, and from SPHERES back to CUBES. PARTICLES from the breaking can be measured or weighed as 1, or as multiples or as fractions of 1. The 1 of a given formula can become 10^x or 10^{-x} in some other formula.
>
> In its ultimate analysis, 1 is the accumulation from $\pi - 3 = Z$ (See PURE WATER and GAS.); and 72 is less than 1:

$$7 \times 0.14159265 = 0.99114855\ldots$$

The 3 is $1 + 2 = 3$ (See DIGITS.). BASE $= 10$, and $10 - 3 = 7$: 2 CUBED $-$ 1 CUBED $= 7$ is the WHOLE NUMBER FRAGMENTATION FORMULA for SINGLES and PAIRS in the first break of CONTINUOUS CHANGE.

Figure 1 is an example of his numerical work.

It is difficult to know what to make of it. There are recognizable difference tables, there are symbols, there are a lot of numbers, and there are words here and there. Some of the words are mathematical—cubes, hypotenuse, fifth power, and so on—but on other sheets there are also words from other sciences: centrifugal force, water density, sun cycle, oxygen, lutetium, and many words from chemistry and physics. There are also, as in the example, lines going here and there for no apparent purpose, and also strange capital letters appearing seemingly at random on the page, with no explanation. The example is typical and is far from the most bizarre of W.'s sheets. All his material is incomprehensible.

One sheet includes a picture of W. He was standing in front of a bookshelf containing more than forty loose-leaf binders, with labels attached—"216," "6000," "Numbers," "Solution," "Valance," and so on, part of the world's most complete mathematics library—holding one of his binders in his hand, and looking straight into the camera. He was a thin man with a light-bulb shaped head, completely bald, with large and protruding ears. His face was lined and his skin was loose; he was probably in his seventies when the picture was taken. He was wearing a double-breasted jacket with astounding lapels protruding almost to his shoulders, not in the style of the year the picture was taken or of any other year I can remember. Underneath his jacket he was wearing no shirt, only an undershirt. His face, very like the man's in Grant Wood's *American Gothic*, was ineffably sad. It is a poignant picture.

W. wanted to make money from his work:

> Your TIME is worth MONEY, if you use your time to MAKE MONEY. My COMPLETE METHOD is the BEST TOOL YOU CAN BUY, to learn HOW TO SAVE ALL OF YOUR TIME FOR MAKING THE SUCCESS THAT WILL BE WORTH MONEY TO YOU. Send only $7.50, today, to make your first investment in keeping ALWAYS AHEAD of all of your competitors.

But he also wanted mathematical acceptance:

> [Last year], I began sending samples of my COMPLETE MATHEMATICS METHOD, to all United States colleges and universities. I chose

x	x^3	d_1		**R**
1	1			
2	8	7	$(1 \times 6) + 1 = 7$	x $\quad x^3 - x^2$ —— d
3	27	19	$(3 \times 6) + 1 = 19$	3 $\quad 27 - 9 \quad = 18$ **A**
6	36	26	$H_3O^{+1} \quad = 19$	**F** $\qquad H_2O = 18$

C

$(d_x = 3$ ——— $d_2 = 18)$ ————————————⌐

H_2O is 3 ATOMS. H_2O WEIGHT = 18. H_2O DENSITY = 1.

x	x^2	d_1	x	x^2		EVEN NUMBER SUM
2	4		20	400	HYPOTENUSE	$2 + 4 + 6 + 8 = 20$
R 7	25	21	21	441	29	$4 \times 5 = 20$
	29		41	841	**D**	$(20 \times 10^3 = $ EQUATOR/2 in kiloMETERS

(Here: **R** 7, 29 etc.)

$7^2 = 49$ —————— $49 - 3^2 = 40$ —————— $40/2 = 20$

x	x^2	d_1	x	x^2	**H**	$39 - 20 = 19$	$(19 - 18 = 1)$
5	25		39	1521	HYPOTENUSE	$80 - 21 = 59$	$59 + 1 = 60$
8	64	39	80	6400	89	$89 - 29 = 60$	**D**
13	89		119	7921	**S**	$89 - 29 = 60$	

(See 6 RELATIONS.).
13, 39, 40, 41, 42

$13^2 = 169$ ————— $169 - 3^2 = 160$ ————— $160/2 = 80$

$x = 3, 4$ ——————— RIGHT TRIANGLE 7, 24, 25

CUBES

EARTH EQUATOR = 24000 MILES

x	x^3	d_1	
24	13824		
25	15625	1801	1801
$7^2 = 49$ 29449			1800

R

1

FIFTH POWER

x	x^5	SUM	d_1	d_2	d_3	d_4	**A**
5	3125						
6	7776	10901					
7	16807	24583	13682				
8	32768	49575	24992	11310			
9	59049	91817	42242	17250	5940		
10	100000	159049	67232	24990	7740	1800	
45	219525	335925	148148	53550	13680		

$(900/10^2 = 9)$ **F**

(45 is DIGIT SUM.). ——————————⌐

FIGURE 1

20 state universities, all the way from Alaska and Hawaii to Florida and Maine; 20 big private universities; and 60 small private and church colleges; for the first mailing. [Later] I mailed my CONTINUOUS CHANGE, from 8, 15, 17 VARIABLES, to 200 DIFFERENCE, to the MATHEMATICS DEPARTMENTS of 250 colleges and universities. Department chairmen of 3 universities wrote to me.

On one of his sheets there is a photocopy of a letter from the head of a department of mathematics of a small college in the southeast:

Dear Sir:

The first envelopes were received on time in very good shape. In fact they were received before the letter. The comments and correction have also been received. We will use these materials as a supplement to my two classes in mathematics for teachers. One student has the material now to be presented to the class as a project. This complementary material is greatly appreciated from me as the teacher and from the class as a whole.

The letter is baffling. It cannot be a forgery, since the letterhead looks genuine and W. was surely unable to write in that style. I cannot believe that the department head was able to make sense out of "1 Relations," and I cannot imagine what the class project could have been. I wrote to the college from which the letter came, but no one there had any memory of W. or his Complete Mathematics Method.

How W. thought his material could be applied to anything is hard to imagine. There was a hint or two of what he had in mind:

WORLD'S MOST COMPLETE MATHEMATICS LIBRARY is KNOWLEDGE for use in ALL INDUSTRIES.

SO NEW competitors have not yet thought of publishing an imitation of it!

At its own SPECIFIC PLACE in TIME and SPACE, each PURE SUBSTANCE has its PERFECT WHOLE NUMBER VALUE. My COMPLETE MATHEMATICS METHOD gives you this PERFECT WHOLE NUMBER; in less time than you can feed a computer, and then read back the answer. You need my COMPLETE MATHEMATICS; within your reach; at all times. The 6 PARTS, indexed for you on this page, are ready for your order, and will be shipped to you immediately after your order reaches me.

A whiff of Pythagorean number mysticism comes to the nostrils. But the odor of insanity also hangs over W.'s work. Only someone crazy would think that difference tables and Pythagorean triangles could be applied to solve real problems.

BASE FOR THE NUMBER SYSTEM, THE BEST

Ask anyone "What is the base of our number system?" and you will get the same response: "What?" However, when you explain, "You know, like 1776 is one thousand, seven hundreds, seven tens, and six ones" you will get, sooner or later, the realization in the person you ask that 10 is the key number. However, the realization will not come as a revelation, since everyone knows, semiconsciously, that our numbers are built on powers of ten. The way we write numbers is as natural as breathing. In the same way, people do not think much about their lungs, though they use them for breathing all the time. Numbers, air: they are just *there*, and everyone assumes that they always have been there, just as they are now.

When people do not think about a system very much, it is a sign that the system is operating very well. There is no higher tribute to an artificial structure than for everyone to think that it is natural. However, our way of representing numbers is anything but natural, as the history of numbers shows. The idea of place value, where the first 7 in 1776 means something different from the second 7 because it is in a different place, is not a natural one. The ancient Egyptians, among the most mathematically advanced of ancient civilizations, did not use place value. They made a mark when they wanted to represent 1, two marks for 2, and so on up to nine marks for 9. They had a different mark for 10, so 44 would be written as four ten-marks and four one-marks. They also had marks for 100 and 1000, and all the marks always had the same meaning wherever they were placed. Roman numerals use the same idea, but with more marks to reduce the amount of writing. The Egyptians would need 1 + 7 + 7 + 6 = 21 marks to write 1776. The Romans needed only 9: MDCCLXXVI. However, M still meant 1000 no matter what place it occupied. The ancient Chinese civilization had a system similar to the Egyptian. Our place-value system, which seems so natural to us, is really quite unnatural.

We will never know who first had the place-value idea, and we will never know why it arose. However, arise it did, millennia ago in ancient Babylonia. The ancient Babylonians did not use base 10, though. They had a mixed system, with separate symbols for 1 and 10 that they used to represent numbers from 1 to 59. After that, place value with base 60 took hold. To represent 61 they wrote two of their symbols for 1 next to each other; 123 would be two 60s and three 1s: || |||. Similarly, | ||| || would represent one 60^2, three 60s, and two 1s, or 3782. Place value makes doing arithmetic convenient. It allows fractions to be represented as decimals, so the Babylonians had no trouble in writing $\sqrt{2}$ accurate to six places. Replacing their marks with our digits, they said

$$\sqrt{2} = 1.\,(24)\,(51)\,(10);$$

that is,

$$\sqrt{2} = 1 + \frac{24}{60} + \frac{51}{(60)^2} + \frac{10}{(60)^3} = 1.414212962\ldots,$$

very close to the truth. In contrast to the Babylonians, the Egyptians and Romans had great difficulty with fractions and avoided them whenever possible. Place-value notation is the reason why Babylonian mathematics was so tremendously superior to the mathematics of any other ancient civilization. It did not keep Babylonian civilization from collapsing, but the idea lived on.

The place-value idea was taken over by the Indians, who made two improvements. The first was to change the base from 60 to 10. "That was sensible of them" is a natural reaction to that, but it was a surprising event. To have the idea is not the surprising part, since ideas are fairly cheap; it is that the idea was taken up by an entire culture. People do not change their bases lightly. One can only speculate that the ancient Indians had some primitive way of representing numbers based on powers of 10, and the obvious superiority of the place-value system made it easy for them to give up the primitive way. The second improvement, which was not made until some time around 500 A.D., was the invention of a symbol for zero, a placeholder with no value. The Babylonians got along without a zero, but there is no denying that 0 is a handy thing to have.

The reason for the choice of ten for a base is close to hand. In fact, it is *on* our hands, namely our fingers. This is so obvious as to require no comment. However, it is a shame that humans were not constructed with six fingers on each hand. Then we would count by dozens and not by tens, in the duodecimal rather than the decimal system, and computational life would be easier. The existence of the word "dozen" shows the need for twelveness in everyday life, as does its many appearances in measurements of all sorts of quantities: a dozen inches to a foot, a dozen months in a year, two dozen hours in a day, and 30 dozen degrees in a circle. A grocer is a person who deals with quantities by the gross—a dozen dozen. Beer comes in half-dozen packs. There are dozens all around us.

Dozens are handy because they are evenly divisible into halves, thirds, fourths, and sixths. Tens, on the other hand, are handy only for halves and fifths, and we want to divide things by 3 and 4 much more often than we want to divide them by 5. The 24-hour day divides into three shifts and the 12-month year into four quarters. A 20-hour day or a 10-month year would not work as well. Decimals also would be much nicer if we counted by dozens. In base 10, 1/3 is

$$0.333333\ldots = \frac{3}{10} + \frac{3}{10^2} + \frac{3}{10^3} + \frac{3}{10^4} + \cdots,$$

whereas in base 12 it would be simply

$$0;4 = \frac{4}{12}.$$

(We need a new symbol for the duodecimal point to avoid confusion with the decimal point, and the semicolon is handy.) Similarly, one-sixth would be 0;2. One-eighth would be even simpler than 0.125:

$$\frac{1}{8} = \frac{1}{12} + \frac{6}{144} = 0;16.$$

Two-thirds (= 8/12) would be 80 percent. Actually, it would be 80 per *gross*, so it would be better to say 80 pergro; but whatever it is called, 80 is more convenient than 66.6667. Counting by twelves has advantages.

These advantages have not gone unnoticed. F. A. (1902–1978) was a zealous promoter of counting by dozens. In "An Excursion in Numbers," an article that appeared in the *Atlantic Monthly* in 1934, A. advocated abandoning the decimal system in favor of duodecimals. As might be expected of a contributor to the *Atlantic*, A. was no crazed fanatic. He was the director of publications of the Russell Sage Foundation for twenty-eight years, 1928–1956, and was the author of *American Foundations for Social Welfare* (1946), *Philanthropic Giving* (1950), *Corporation Giving* (1952), *Philanthropic Foundations* (1956), and several other books of a similar nature. He also wrote fiction (*Grugan's God* (1955)), popular mathematics (*Numbers, Please* (1961)), and even in retirement he could not stop producing books (*The Tenafly Public Library: A History* (1970)). In 1952 his alma mater, Franklin and Marshall College, thought enough of him to make him one of its Doctors of Humane Letters.

As A. pointed out, when we count as if we had six fingers on each hand we have two extra digits and so we need two new symbols for them. The digits in his system are

$$1, 2, 3, 4, 5, 6, 7, 8, 9, \chi, \text{ and } \varepsilon.$$

(the last two are pronounced "dek" and "el"). The next counting number would be 10: one dozen and no ones. 100 would be a gross, duodecimal 1000 would be $12^3 = 1728$ in decimal, and so on. If we wanted to, we could call 10 "do" (for dozen, but with a long *o*) and 100 "gro," so 245 could be said "two gro four do 5" or, after everyone had changed over to counting by twelves, "two forty-five" just as now. Everyone would have to learn a new addition and multiplication table, but it would not take long until $9 + 5 = 12$ and $7 \times 8 = 46$ became as natural as their decimal counterparts, and children would be whipping off problems like

$$
\begin{array}{r}
76 \\
87 \\
125 \\
+\ 59 \\
\hline
303
\end{array}
\qquad
\begin{array}{r}
216 \\
\times\ 73 \\
\hline
646 \\
12\chi6 \\
\hline
134\chi6
\end{array}
$$

just as fast as they do them now in base ten. People who were too old to change or who could not master the new tables could be issued duodecimal calculators. After one generation at most, the transformation would be complete and everyone would look back with pity at their ancestors who had to struggle along in the inconvenient base χ system.

Of course, we are not going to change. As A. himself wrote,

> The unhappy fact remains that man is ruled more by habit than by reason. We shall continue counting on our fingers in the logically silly system of 10 to the end of our days.

Nevertheless, the idea was taken up, with more or less seriousness, by other people. For a time there was a Duodecimal Society of America, with members, publications, an official journal, *The Duodecimal Bulletin*, and translations into Esperanto: *Ekskurso en Nombroj* and *Antipatio al Arithmetiko*. Of all languages to translate duodecimal literature into, Esperanto is the most appropriate, since an esperantist is almost exactly equivalent to a duodecimalist: both are advocating changes that are logical, would be beneficial, and are not going to occur.

Interest in duodecimals did not go away. Inspired by A., G. T. wrote *The Dozen System*, subtitled "An Easier Method of Arithmetic," a 56-page book (56 means 5 dozens and 6 ones, of course) published by Longmans, Green in 1941. Most publishers of works that lie on the fringes of mathematics are on the fringes of publishing—vanity presses, or houses that specialize in the occult—but Longmans, Green was not a publisher of that kind. That Longmans published the book is a measure of the respectability that duodecimals had at the time.

The book is a sober and unpretentious work. The farthest it goes into frivolity is the anonymous verse

An Arab invented a symbol for nought;
    Roman ridicule was his lot.
T'was later adopted without further thought.
    This is as far as we've got.

The grocer who had to wrap his goods
    Has counted by the gross since time began.
His dozen divided up several ways;
    He was a practical man.

A Frenchman divided the metrical Earth,
    Unsymmetrical ten he chose;
But he missed half the force of the decimal point,
    Still counting on fingers and toes.

So our multiplication continues absurd
    With its clumsy quarters and thirds.
By dozens it's better; means less work in school;
    But what is the use of words?

T. mentioned the possibility of using a base other than 10, giving 6, 8, 12, and 16 as examples, and modestly continued:

It is not to be expected that the world will quickly adopt another method of counting. So much emphasis has been given the number Ten that it appears to have qualities almost magical. Other numerical bases have been so neglected that no one knows what simplifications will or will not result from their use. Is it not worth giving a little thought to a system which simplifies arithmetic, without too much concern as to what may or may not happen in the distant future?

He set out the advantages of using 12 as a base:

A. It contains more different factors, that is, it splits up in more different ways than any of the others.

B. It is the smallest number with four divisors.

C. It has more simplifications to offer to arithmetic than any of the others.

Which base is best? Six is convenient, he said,

> eight is less good, though perhaps better than ten. Twelve is the best; and sixteen, little better than eight.

He went on to explain how to add, subtract, multiply, and divide using duodecimals. He pointed out that it is easier to recognize divisors in base 12:

Tests for divisibility:—

by	2	Last digit even.
	3	Last digit 0, 3, 6, 9.
	4	Last digit 0, 4, 8.
	6	Last digit 0, 6.
	8	Last two digits divisible by 8.
	9	Last two digits divisible by 9.
	ε	Sum of digits divisible by ε.
	10	Last digit 0.

He mentioned interesting facts, such as that all Mersenne primes ($2^p - 1$ with p prime) greater than 7 end in 27 or $\chi 7$, and that all even perfect numbers ($2^{p-1}(2^p - 1)$) greater than 24 end in 54. He gave some arithmetical problems:

1. In how many ways can unity be expressed as the sum of two fractions using each digit once?

$$\text{e.g. } \frac{136}{270} + \frac{48\chi}{95\varepsilon} = 1$$

Ans. About five dozen (Decimally about one dozen.)

2. How many numbers from one up can be expressed with four fours using the four arithmetic processes and duodecimals?

$$\text{e.g. } 26 = \frac{4 - .\dot{4}}{.4(.\dot{4})}$$

Ans. Two and one half dozen. (Decimally twenty two.)

3. How many numbers from one up can be similarly expressed with the four digits 1, 2, 3, 4 each used once?

$$\text{e.g. } 99 = \frac{32 + 1}{.4}$$

Ans. Nine dozen and nine. (Decimally eighty eight.)

His duodecimal system of weights and measures is based on the palm, one-twelfth of a yard. A pound is the weight of a cubic palm of water (at maximum density), and a pint is its volume. Thus we have

Linear measure	Weight
10 Quins = 1 Palm	10 Ounces = 1 Pound
10 Palms = 1 Yard	1000 Pounds = 1 Ton
1000 Yards = 1 Mile	

Volume

10 Ounces = 1 Pint
10 Pints   = 1 Gallon
10 Gallons = 1 Bushel or Barrel
10 Bushels = 1 Cubic yard

All those 10s are of course dozens, so the duodecimal mile is 1728/1760 = 98.2% of a decimal mile, and the difference would hardly be noticed.

The Dozen System is an admirable book. However, it lacked the fire necessary to ignite the flames of revolution. What would have been necessary was something that would incite ordinary citizens to take to the streets, demonstrating en masse against the tyranny of base ten. It is difficult to think of what that could have been.

You would think that World War II would have taken peoples' minds off duodecimals, and that the Society would have been one of the casualties of the war, but it was not so. The Duodecimal Society of America was incorporated in 1944 and its journal, *The Duodecimal Bulletin*, was first published in 1945. A British Duodecimal Society was founded in 1958. The American Society, as part of a membership drive, published its *Manual of the Dozen System* in 1960. The Society said that

it is due in large measure to the work of the Duodecimal Society, that the faults and limitation of the decimal system are now widely recognized.

Also,

Twelve, then, appears to remain as the ultimate choice as the base of the number system. It can efficiently and comfortably accommodate the entire field of measurement to obvious advantage.

The Society was hopeful, though realistic:

> The problem of effecting the change in the accepted radix is as formidable as ever. Yet we are more accustomed to change. The amazing scientific developments of today have led to an intense scrutiny of our educational system. Education of the public in the use of the twelve-base may receive added impetus because of this revaluation.

However, the Society may have been going too far with its system of duodecimal weights and measures. Instead of T.'s familiar pounds and pints, the Society's system contained strange new units: a dozen karls is one quan, a dozen quans is one palm, a dozen dribs make one dram, a dozen drams make one founce, a dozen grovics equal one minette, a dozen minettes equal one temin, a dozen temins make one duor, and so on. Those units have not caught on.

The original Duodecimal Society disappeared some time after 1960, but it has been revived as The Dozenal Society of America, and *The Duodecimal Bulletin* continues to be published. χ and ε have had to adapt to modern typography:

> The DSA does NOT endorse any particular symbols for the digits ten and eleven. For uniformity in publications we use the asterisk (∗) for ten and the octothorpe (#) for eleven. Years ago, as you can see from our seal, we used χ and ε. Both χ and ∗ are pronounced "dek". The symbols # and ε are pronounced "el".

The Fall 1195 issue of the *Bulletin* has an interesting problem: find dozenal numbers (the Society prefers "dozenal" to "duodecimal") that are exactly twice as large as their decimal counterparts. One answer is 11788. Quite right:

$$1 \cdot 12^4 + 1 \cdot 12^3 + 7 \cdot 12^3 + 8 \cdot 12 + 8 = 23576 = 2 \cdot 11788.$$

Other solutions, the *Bulletin* says, are

> 11790, 11818, 11820, 12298, 12328, 12330, 24658,
> 24660, 25200, 25168, 25170, 36968, 36990, 37528,
> 37530, 38038, 38040, 49858, 49860.

Are those *all* of the solutions? What about dozenals that are three or four or five times as large as base-ten numbers with the same digits? Dozenals can raise interesting questions.

The issue contains good-humored filler:

> Don't throw this *BULLETIN* away—
> Give it to a friend, or
> Leave it in your dentist's office

It also contains simple number theory (a long article whose climax is the discovery that perfect numbers, when written in base 2, consist of a string of 1s followed by a string of 0s), scholarship (an article by Anton Glaser, author of *History of Binary and Other Nondecimal Notation*, showing that the belief, commonly held in some circles, that Charles XII of Sweden (1682–1718) was an advocate of the duodecimal system is in error), and a list of publications available from the Society:

> Our brochure
> "An Excursion in Numbers" by [F. A.]
> *Manual of the Dozen System* by [G. T.]
> *New Numbers* by [F. A.]
> *Douze: Notre Dix Futur* by [J. E.]
> Dozenal Slide rule, designed by [T. L.]
> Back issues of the *Duodecimal Bulletin* (as available)

The list shows that the duodecimal movement has not strayed far from its foundations. The issue's cover has a group picture of some of the attendees at the Society's annual meeting in 1194. The Society was not thinking clearly when it published the picture: there are *eleven* people in it. Why not a dozen?

Base 12 is not the only possible replacement for base 10. E. T. published a small pamphlet, *Calculate by Eights, Not by Tens*, in 1936. The thirties were evidently a time when the idea of changing bases was in the air. T. said that halves are much more natural than fifths:

> Our minds prefer the simply easy even operations on things, repeated many times where necessary. We divide things evenly into halves with our eyes. We judge equality of weights with our two hands. We split differences to close trades. Surely these are facts, not opinions.
>
> We like to halve and double things. But we are now forced by our decimal arithmetic to calculate about these even halves and doubles with fifths and odd and prime fives.

True enough, but the inference he drew does not follow:

> Our minds and the nature of things will never change in respect to halves and doubles. We cannot change our minds to fit out present decimal arithmetic. We must adapt our arithmetic to our minds. Therefore we must correct our arithmetic and measures from scale ten to scale eight.
>
> The old proverb about the fit of a square peg in a round hole must be revised to include the lack of fit of our five-sided arithmetic in our foursquare minds.

As pairs are favored in our minds and in all nature our arithmetic must favor pairs. By using the base eight, the product of pairs, our calculating system will best correspond with many features of nature.

He recognized that change would not be easy, and called for help:

If a change of base is assumed impossible where before did an impossibility deter a real scientist from study of a subject and insistence in his conclusions? Psychologists, that vast influential and international group, students of the mind, should assert their authority that eight is the best base for arithmetic.

Even though in base 8 the day would have 30 hours, nicely dividable into three 10-hour shifts, thirds would be no easier than they are in base 10 $(1/3 = .252525 \ldots)$, and the other advantages pointed out by T.—even steps in denominations of money, for example, in place of the uneven sequence $1, $5, $10, $20, $50, and so on—do not seem all that important. Even so, he had his system published in *School Science and Mathematics*, and prominently featured that fact in his later output. Here is an example, dated January 37, 3616:

Who can make these troublesome thirds and fifths accurately without mechanical aid? Fifths, difficult for our hands and minds, are always present in our decimal and metric system. Psychologists should measure and declare our eternal preference for halves above thirds or fifths, but they neglect the great opportunity. Psychologists should condemn the metric system with its confusing fifths. Arithmeticians and educators wrongly favor the scale of twelve containing in the place of five the *equally offensive* odd and prime three.

Yes, halves are natural and permeate our weights and measures, with the inch divided into sixteenths, 16 ounces in a pound, four quarts to the gallon, and so on. However, people are not sufficiently offended by the odd and prime five to abandon its part in our number system. (The idea that primes could be offensive was new to me. Austere, perhaps, in their refusal to be broken up into parts, but hardly *offensive*.)

Yet another system was advocated in *Letter Systems in Business and Technology*, a neatly printed pamphlet written by W. K. and printed in 1942. In it, K. said that we should replace our base-10 numerals with ones in base 24 *and* base 60. The digits in base 24 would be the letters from a to w so, as with the ancient Greeks, every word would also be a number (except for words with letters past w in them, like slyly, sizzle, and syzygy), and the science and art of gematria could flourish anew. The digits in base 60 would be the letters from a to v, omitting i and j, repeated three times with primes and double primes, so a to v would represent

1 to 20, a' to v' 21 to 40, and a'' to v'' 41 to 60. Thus the value of the word "one" would be

$$13 \cdot (60)^2 + 12 \cdot 60 + 5 = 47525,$$

and $2000 = 30 \cdot 60 + 20$ would be written $o'v$. Large numbers would be represented with fewer digits, as

$$\text{don}'t = 4 \cdot (60)^3 + 13 \cdot (60)^2 + 32 \cdot (60) + 18 = 912738,$$

which, K. said, would be a great savings:

> In salvage for war production letter numbers for auto tags may save many dollars and many tons of iron and paint.

As for the difficulty of memorizing a 60-by-60 multiplication table—something even the ancient Babylonians did not insist on—K. said

> The letter multiplication table would be no more difficult to learn in early life than to learn to spell the ten thousand words which constitute our working English vocabulary. Furthermore, we are already segregating bright pupils from the ordinary. These pupils could absorb the letter table and become mathematical wonders to the rest of the population. Such an achievement would be of infinite value to those who later take up accounting, technology or science.

Besides the advantages of more compact notation and having more decimals (sexagesimals, they would be) come out evenly,

> As to practical applications of letter systems they may be used commercially in saving time, space and material. Where customers have to quote a quantity the 24 system is best, tho it does not save as much as the 60 system. We believe, however that the 60 system could be used for auto tags, since they are mainly used for the police who could be trained therein. Officers can memorize three letter digits better than six figures. Hotels and large department enterprises can save several hundred dollars in letter numbering.

For some reason,

> We anticipate violent opposition by publishers to a book on the subject; so, if you would buy such a book, would you kindly write us, so that we may have some ammunition for the battle.

More likely than "violent opposition" would be "bored indifference."

Even though we will never change bases, the base-changers are right about base ten not being the best. It has been gratifying to see the failure of the efforts to metricize the country. Some groups pushed very hard to make base ten universal by eliminating feet, pounds, and degrees Fahrenheit from the United States in favor of meters, kilograms, and degrees Celsius, but they failed. Their failure was partly owing to unreasoning resistance to change, but I like to think that another reason was the realization that our system of weights and measures, with its quantities divisible by 2, 3, and 4 (and not by 5), has, in its evolution over time, come to fit the needs of humanity better than an artificial system thought up by the French two hundred years ago.

We are stuck with base 10 until the collapse of civilization makes a fresh start possible, and even then those making the fresh start would no doubt count on their fingers and so bring us back to base 10 again. Today, change is even less likely than it was in the past, since the main argument of the reformers—that calculations with numbers would be simpler and quicker in their new base, thus saving time—has, because of pocket calculators, lost most of its force. Thus does tradition triumph over logic once again.

Nevertheless, the Dozenal Society carries on, good-humoredly fighting a battle that cannot be won. The annual dues are a mere dozen dollars a year; the Society's address can be found in the Notes.

BITTERNESS, CRANKS'

If you are a crank, what are you to do? You send out your material, over and over, and people do not read it, or they read it and fail to understand what is perfectly clear to you; you are brushed off and you do not get the recognition that you so richly merit. You may start to think there is a conspiracy to keep you down. Here is what G. G., an amateur who made some observations on numbers, wrote in response to a suggestion that he show his material to someone in the mathematics department at the state university of his state:

> The University of [C.] did not have a strong mathematics department when I was there, so I decided to send my manuscripts somewhere else. I may try to send my manuscripts to the math journals, but after the delay occasioned by my failure to find one that would evaluate it instead of rejecting it, I decided to try sending them to the Universities. I have written to Princeton and Yale, but haven't heard from them yet. How fortunate Ramanujan was that he didn't have to try to publish a manuscript in America!

How odd that G. did not see that rejecting his manuscript *was* an evaluation of it. In another letter he wrote:

> Because of your not replying I have lost a year that I could have used either in getting a good position in industry or in continuing my education; I had hoped to use the recognition that I would have received to one or the other purpose. Apparently there is no room in the American mathematics establishment for any except its chosen few.

Being rejected, being ignored: it goes on and on. What can you do? One thing you can do is stop trying. This often happens, and I expect that it happened with G.; but it is unremarked on since former cranks do not send out letters to former correspondents saying "I've stopped now." Another thing you can do is carry on, ignoring all rebuffs, secure in the conviction that you are right, everyone else is wrong, and vindication is just around the corner. This also often happens. However, if the strength of your ego (or the extent of your delusion) is not up to all of the battering it can take, the corrosive effect of time produces only increasing frustration and bitterness. Stopping, madness, or bitterness: there seem to be no alternatives.

I have two samples of the work of I. I. Here he is in 1984, describing his work on what he called the Grand Unified Field of mathematics:

> This will turn out to be the greatest discovery in mathematics in many centuries. There is no doubt that the Grand Unified Field is now in our hands. And it will drastically change our overall approach to mathematics beginning with our instruction in elementary school. This is the ultimate in mathematics. Every student, every mathematician, every scientist, and every engineer will find some application of the subject to his or her favorite area. . . . It is not algebra, not geometry, not statistics, not calculus, not number theory, or any other classification. It is all of them together and covers every field of mathematics, including some which have not yet been invented. . . . In the Grand Unified Field, all mathematics is brought together and interconnected. . . . It was the Grand Unified Field which Lewis Carroll (Charles Lutwidge Dodgson) was trying to describe, but he did not have the map, to guide Alice about in Wonderland. . . . Entering the Grand Unified Field, one must leave the known world of higher mathematics and enter a new world where rigor is absolute. Many other mathematicians have entered the field but had no map and failed to locate sufficient landmarks to relate it to conventional mathematics. Among these are Euler, Gauss, Napier, and many others.

Sweeping claims, exuberantly made. One is filled with curiosity about what the GUF could possibly be.

It turns out to be nothing much, unless I. was keeping something back. Take four integers (b, e, d, a) such that $b + e = d$ and $e + d = a$, such as $(1, 1, 2, 3)$ or $(5, 2, 7, 9)$—any start of a Fibonacci sequence, that is—and let

$$A = a^2, \quad B = b^2, \quad C = 2de, \quad D = d^2,$$
$$E = e^2, \quad F = ab, \quad G = d^2 + e^2, \quad H = 2de + ab,$$
$$I = |2de - ab|, \quad J = bd, \quad K = ad, \quad L = \frac{beda}{6}.$$

You thus have twelve numbers to play with, or sixteen if you count the starting values. From $(1, 1, 2, 3)$ results

$$(A, B, C, D, E, F, G, H, I, J, K, L) =$$

$$(9, 1, 4, 4, 1, 3, 5, 7, 1, 2, 6, 1),$$

in which $(C, F, G) = (4, 3, 5)$, the sides of a right triangle. From $(5, 2, 7, 9)$ comes

$$(81, 25, 28, 49, 4, 45, 53, 73, 17, 35, 63, 105)$$

and $(C, F, G) = (28, 45, 53)$, the sides of another right triangle. This is not surprising, since no matter what a, b, c, and d are,

$$C^2 + F^2 = (2de)^2 + (ab)^2 = 4d^2e^2 + ((d + e)(d - e))^2$$

$$= 4d^2e^2 + d^4 - 2d^2e^2 + e^4 = (d^2 + e^2)^2 = G^2.$$

I. discovered also that if you draw concentric circles with radii $C - F$, G, and $C + F$, the outside ring and the inside ring have equal areas. If $(C, F, G) = (4, 3, 5)$, the circles are as in Figure 2. The area of the outside ring is $49\pi - 25\pi = 24\pi$, the same as the area of the inside ring, $25\pi - \pi = 24\pi$. This is no surprise either since $G^2 = C^2 + F^2$, so the area of the outside ring is

$$\pi(C + F)^2 - \pi G^2 = \pi(C^2 + 2CF + F^2 - (C^2 + F^2)) = 2\pi CF$$

while the area of the inside ring is

$$\pi G^2 - \pi(C - F)^2 = \pi(C^2 + F^2 - (C^2 - 2CF + F^2)) = 2\pi CF.$$

Also, if you take an equilateral triangle with side $d(a + e)$, then a line with length $d^2 + ae$ drawn from one vertex to the opposite side will divide the opposite

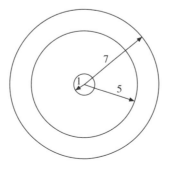

FIGURE 2

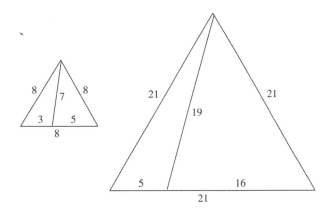

FIGURE **3**

side into two parts, ab and $e(d+a)$, which produces some nice triangles with integer sides (Figure 3). To verify that this always works is probably a pleasant exercise.

That is about all. There is an ellipse or two and the identity

$$\left(\frac{1}{2}\right)^3 + \left(\frac{2}{3}\right)^3 + \left(\frac{5}{6}\right)^3 = 1,$$

and then the GUF peters out into a bunch of numbers and a last paragraph:

> It would seem, with this shaking of the foundations of mathematics, that all mathematics would be voided. But this is not the case. All of current mathematics fits within the fabric of the Grand Unified Field and each part requires only a slight modification, as the Fibonacci numbers were modified. Then each of these parts becomes one of the parts of the Grand Unified Field of Mathematics.

Just as I.'s discoveries are no surprise, it is no surprise that the GUF was not received as I. thought it should be. Here he is again, three years later, in 1987. See what three years of frustration can do:

> On the Grand Unified Field, I can tell you that you can never construct it with conventional mathematics because there are basic errors in the teaching of elementary mathematics, and errors and omissions in definitions from then onward. The mathematics which [you] are peddling holds no future. . . .
>
> Why do you think that the public is completely turned off by mathematics? It is because even a child knows more mathematics than you can teach. You eradicate this knowledge, creating first a mental block and then you try to instill a purely invented mathematics. . . . The "wave theory"

you teach is a total fabrication. Your ellipses are only theory, and wrong theory at that. Totally wrong. . . .

So, right now you are saying, "What is this? Some kind of kook?". That may be your opinion and you are entitled to it. But can you turn any six-place decimal into a prime fraction? The Grand Unified Field can do it in ten seconds. Can you solve any problem with ten unknowns with only two equations? The Grand Unified Field can in most cases. Never are more than three equations required. Can you solve "Plato's Disc of Gold"? The most powerful computer is of no use without knowing the Grand Unified Field. But any high-school student can solve it with a pencil and a single sheet of paper, if that student knows the Grand Unified Field. And that student will know more mathematics than you can teach him in twenty years.

Do you know what a Gaussian integer is? You think you do, but it is wrong. I know what it is and where it comes from. There is no such thing as -1^2. Do you know what a "quaternion" is? I do, and so did Plato. A Gaussian integer is a combination of five quaternions, involving integers, squares, and gnomons. . . .

You've had your chance in promoting the Grand Unified Field. . . . So long as you know *all there is to know* about mathematics, there is no point in going further. . . . Conventional math can never break through into the Grand Unified Field because it is necessary to start right at the beginning, as I started 40 years ago. . . .

If you want to know how to reach the Grand Unified Field through me, . . . I want $1000 per page (double spaced, of course) paid in advance, and then only a maximum of three pages.

Anger, red hot, and bitterness, most sour. Thus time does its work.

Here is "Plato's Disc of Gold" problem, mentioned above.

Dear Archimedes:

Your problem is solved but:— About twenty years ago he lived on Crete and was moving to Thrinacia. The cost of transporting his herd to Thrinacia was one ounce of gold per animal with all shiploads being equal.

His gold was in large circular discs with a hole one unit in diameter at the center. Each disc was uniform in thickness and all had the same weight, but they varied widely in diameter. He divided the discs into concentric rings having whole numbers for their weights and inside and outside diameters. A successive outer and inner ring from any disc would pay for each shipload. One-half of the weight of the outer fifteen rings was in the outer six. Again dividing the outer six, leaves the outer five balancing the next six. How many discs were there, to use the last disc of gold? How many animals per load, and how many loads, if the number of discs is a triangular number, the number per load is a square number, and the number of loads is a product of a pentagonal number, a square number, a triangular number and of a cubic number?

s/s Eratosthenes

The problem is probably very hard to solve, with or without a Grand Unified Field.

CALCULUS, CELESTIAL

This is the name given by its author, H. J., to his discoveries in mathematics that he hoped would have applications to physics:

> It may be too that we now have the means with which to investigate the Unified Field Theory of a set of universal laws encompassing not only gravitational and electromagnetic fields but also those associated with the atom.

He presented his work in a 34-page pamphlet in 1986.

> Around 1947, my detection of restraints common to derivatives of $f_1(x) + f_2(x) + \cdots + f_n(x)$ and $f_1(x)f_2(x) \ldots f_n(x)$ at x seemed hampered by the customary differentials δx and $\delta f(x)$. On recalling that $x = x \cdot 1$ and $f(x) = f(x) \cdot 1$, I let $x \cdot 1$ increase to $x \cdot u$ and thereby cause $f(x) \cdot 1$ to increase to $f(x) \cdot v$, giving

$$\underset{u \to 1}{\text{limit}} \frac{\delta f(x)}{\delta x} = \frac{f(x)}{x} \left(\underset{v \to 1}{\text{limit}} \frac{u-1}{v-1} \right)$$

That is a different way of looking at how a function changes: instead of adding a little bit to x and seeing how much is added to $f(x)$, as is done in all calculus textbooks, *multiply x* by something a little bit bigger than 1 and see what $f(x)$ is multiplied by. There is nothing cranky in that idea, but there is in J.'s estimation of its importance:

My astonishment and elation on finding myself producing what is clearly the most unifying and comprehensive identity in mathematics can well be imagined.

That grand J.'s idea is not. Nice, however, it is. All students of calculus spend a lot of time looking at derivatives, $D(y) = dy/dx$. J. looked at

$$L(y) = \frac{d(\ln y)}{d(\ln x)}$$

instead. This new operator has several pretty properties:

$$L(x^n) = n, \qquad L(e^x) = x, \qquad L(u^n) = nL(u),$$
$$L(uv) = L(u) + L(v), \qquad L\left(\frac{u}{v}\right) = L(u) - L(v).$$

On the other hand,

$$L(u + v) = \frac{uL(u) + vL(v)}{u + v}$$

and $L(Ly)$ is a mess, but one cannot have everything. It might be great fun to develop the theory of the L-derivative and the L-integral and see the parallels with ordinary derivatives and integrals. What is true when $L(y) = 0$? When $L(y) > 0$? For what y is $L(y) = y$? What is the physical meaning of an L-integral? Such questions are worthy of undergraduate research, or at least of being put among the hard problems in a calculus book.

CANTOR'S DIAGONAL PROCESS

The infinite can boggle the mind. The infinite *has* boggled minds for thousands of years. The reason is obvious enough: the infinite is so...so *large*. However, some minds refuse to stay boggled and try to come to grips with something that is essentially ungrippable. They meet with varying degrees of success: some get a handle on some part of the idea, while others slip off into crankery or worse.

Georg Cantor (1845–1918) did both. He founded modern set theory with genuinely new ideas, but he ended his days in a mental hospital. He showed there are exactly as many quotients of positive integers as there are positive integers (the pairing

$$\frac{n}{m} \leftrightarrow n + \frac{(m+n-1)(m+n-2)}{2}$$

matches up the elements of the two sets, with none left out and none left over), and his diagonal process demonstrates there are more irrational numbers than positive integers by showing there is no way of pairing up the two collections. If you try, as with

$$1 - .\underline{7}3890561\ldots$$
$$2 - .5\underline{4}598150\ldots$$
$$3 - .29\underline{8}09580\ldots$$
$$4 - .888\underline{6}1105\ldots$$
$$5 - .7896\underline{2}960\ldots$$
$$6 - .62351\underline{4}91\ldots$$
$$7 - .388770\underline{8}4\ldots$$
$$8 - .1511427\underline{7}\ldots$$
$$\cdots$$

then, to get a number that is not anywhere on the list, make one so that its first digit is different from 7, its second digit is different from 4, its third digit is different from 8, and so on down the diagonal. One such number (there are infinitely many) is the one with digits each one less than the digits down the diagonal, .63751376. . . . It is not on the list, anywhere, because it differs in at least one digit from every number on the list. It is not the first number on the list because its first digit differs from 7, it is not the second number on the list because its second digit differs from 4, and it is not the 3,141,563rd number on the list because its 3,141,563rd digit differs from whatever is the 3,141,563rd digit of the 3,141,563rd number on the list.

Cantor's diagonal process is indeed wonderful. There are no terms to be memorized, no technique to be mastered; it is pure *idea*, and it leads to a lovely structure of infinities, far transcending the finite universe. The process is not only wonderful, it is useful for many things. It can be used to show, elegantly, that a number of things are impossible, for example that it is impossible to write a computer program to solve the problem of determining if a computer program will halt or run forever.

However, some cranks find Cantor's diagonal process unacceptable. Perhaps it is because it shows there is more than one kind of infinity; cranks tend to prefer a simple world. You might think that something as clear (after it is understood) as the diagonal process would allow of no controversy, but the minds of cranks can fail to see almost anything. A 1978 writer found *five* mistakes in Cantor's proof. One was:

> Mistake 4. Cantor's newly discovered number may only be elusive, not external. Suppose we began at the top of the list using the diagonal method and propose to write down 7 for every digit which is not 7 and 0 for every digit that is 7. Such a process would yield 777. . . . It appears not to be on the roster but it is really a number which is always further down the list, which we do not reach because of the sheer expansiveness of the list and the ever-expanding nature of infinity.

The author seemed to view the list as in a state of flux, roiling and seething down toward its bottom, making it impossible to see what was going on there. And when we get there, the seething has not stopped but only receded further, forever out of reach. Infinity is a really *big* idea, and it has strained minds since the time of Zeno.

Here is another of Cantor's errors:

> Mistake 5. The diagonal method is offered without justification. Instructions indicate that given a list, we shall proceed to manufacture a number from this list. At what rate? By what action are we able to overtake "n"? The value of ratios approaching infinity are determinable, yet Cantor's method does not appeal to any reasoning of comparative lengths. On the other hand, if we move to exceed the horizontal length, then why not exceed the vertical (i.e., $n + 1$)? It, too, is ever-expanding. The proof

lacks a reference to length and duration. We thereby fail to gain a perspective on the accuracy of Cantor's method.

That brings to mind a well-known thought experiment. Put two balls numbered 1 and 2 into an urn at one minute to midnight. At $1/2$ minute to midnight, take out ball number 1 and put in balls 3 and 4. At $1/4$ minute to midnight, take out ball 2 and put in balls 5 and 6. Continue putting in two balls and taking out one at $1/8$, $1/16$, $1/32, \ldots$ minutes to midnight. How many balls are in the urn at midnight? On the one hand, it is empty, because if you name a number, the ball with that number was removed from the urn before midnight—you can even determine at exactly what time. On the other hand, at $1/2^n$ minutes to midnight there are n balls in the urn, and that number goes to infinity, not zero, as n increases. One conclusion that can be drawn from this paradox is that it is impossible to do infinitely many things in a finite time. In particular, we cannot pick infinitely many digits in a real number in a finite time. Mistake 5 might have had its origin is some such line of thought.

On the other hand, it might not. It takes a fairly clear head to come to that conclusion, and the author's Mistake 1 is not altogether clear:

> Not every list can be translated accurately into the decimal system. There is nothing magical or universal about decimals; it is just a system whose values depend upon the number 10 and in which many fractions do not find complete expression ($1/3$, $1/7$, $2/11$, etc.). We normally ignore such difference as $1/3 - .333\ldots$ in finite matters, but such ignorance is inexcusable when dealing with infinite matters.

Mistakes 2 and 3 are as impenetrable. The author did not explore any of the consequences of his conclusion that the real numbers are countable.

G. P., the author of *Disproving Cantor's Theory of Transfinite Numbers*, was a holder of the Ph.D. degree. That may be only slightly surprising, since cranks can be any sort of people, but P.'s degree was in *mathematics*. Nevertheless, he produced twenty mimeographed pages arguing against Cantor's transfinite set theory. P. evidently could not stand the counter-intuitive consequences of Cantor's reasoning. To get rid of the result that there are exactly as many positive even integers as there are even and odd integers together (because each n can be matched with its corresponding $2n$) P. decreed that numbers are sets, so that 4 is really a set of four dots $\{\ldots\}$ and 8 is $\{\ldots\ldots\}$. Thus 4 does not match with 8, since 8 has more dots than 4 does.

Another of P.'s arguments was:

> On the basis of our method of 1:1 correspondence we would have to say that these two sets are equivalent. If this is the case then what will we say about the following sets?

1	2	3	4	5	6	7	8	9	...
	2	4	6	8	10	12	14	16	...

We would have to say that these two sets are not equivalent because they are not matched. But these are exactly the same two sets as the first two sets which we declared to be equivalent on the basis of the 1:1 correspondence. No elements have been taken out of any one of the last two sets to make them different from the first two sets. The only difference is that these two sets are not matched. We therefore have here the paradox of having two sets equivalent and not equivalent at the same time. Either we have to say that these last two sets are not equivalent, which is not true, or we have to reject the principle of 1:1 correspondence.

That is reasoning good enough for anyone who does not understand the definition of when sets have equal numbers of elements.

P. revised the definition of equality of sets, so that

1. Two sets can be equivalent and unequal: $\{1, 2, 3\}, \{2, 4, 6\}$
2. Two sets can be equivalent and equal: $\{2, 4, 6\}, \{1, 3, 8\}$
3. Two sets can be nonequivalent but equal: $\{1, 3, 4\}, \{2, 6\}$
4. Two sets can be nonequivalent and unequal: $\{1, 3, 4\}, \{2, 5\}$.

Two sets of integers are equal, you see, if their elements have the same sum. There are probably quite a number of new theorems that could be proved with the new definition.

It is hard to grasp that, when dealing with the infinite, the whole can be equal to some of its parts. P. denied it:

> Does a small line segment have as many points on it as a larger line segment? Answer: No. Reason: We can take the smaller line segment and superimpose it upon the large line segment. This shows the smaller line segment to be equal to a subset of the larger segment. Since a proper subset cannot be compared with the set itself, therefore the two line segments cannot be said to have the same number of points.

Teachers of set theory should be prepared to answer arguments like that.

Another way to deal with the infinite is to deny its right to existence. There are, after all, no physical infinities, so is it not a waste of time, analogous to asking how many angels can inhabit the head—or perhaps the point—of a pin, to talk about mathematical infinities, worry about them, and prove theorems about them? An author writing against infinity said:

The infinite is neither definite or distinct, and neither can it be perceived or conceived by any finite mind. The expression "transfinite number" is also a contradiction of terms since a number really represents something that is countable, meaning by an individual, and must therefore have a finite limit. Outside of such a possibility the infinite must be considered more or less on a par with imaginary values. Since the infinite is uncountable it cannot be given the qualification of a number. The infinite represents something irrational, and only the rational is denumerable.

The set of integers is rational as far as it is countable. However, when we speak of the set of integers as an infinite set, it then becomes an uncountable set, for we are making it go from a rational size to an irrational size, and the whole question of the domain in which the rational ends and the irrational begins cannot be answered by anyone. The basic problem regarding infinity arises when we think of it as a size. Human beings can count numbers but they cannot count them to infinity. Since the counting of numbers is a process that demands time, we can say with definiteness that counting infinitely is a process which demands an infinite amount of time. This shows us clearly that infinity is not a size but a process which is ever ongoing without limit. As such it cannot be reduced to a mathematical symbol which represents a certain size.

Another author, W. D., was not so radical as to deny the existence of transfinite numbers; he merely refused to accept Cantor's proof that there is more than one of them. You do not expect to find Cantor refuted in the pages of a scientific journal, but that is where D. managed to have his 12-page paper, "A Correction in Set Theory," published: in the *Transactions of the [X.] Academy of Sciences, Arts and Letters*. "X" replaces the name of one of our United States.

State academies of this sort, sometimes restricted to science, sprung up in the nineteenth century and helped sustain the intellectual life on the frontier. The frontier of the intellectual life, that is, which moved west much more slowly than the physical frontier. In the nineteenth century they served a purpose, since a college might have only one Professor of Natural Philosophy, expected to provide instruction in physics, chemistry, mathematics, and, from time to time, biology; he might get lonely, and could use the chance to meet with his colleagues once a year to find out the latest developments in the teaching of science. Now state academies have largely outlived their purpose, since there are national and state organizations in each of the scientific disciplines, and it is to the meetings of those organizations that teachers go. But, as institutions tend to do, they have lived on—after all, they do no one any harm and cost very little, so there is no incentive for doing away with them—but they no longer attract material of the highest quality. In some states, manuscripts are not sent to referees, and the editor of the academy's *Transactions*, perhaps with an assistant, decides to accept or reject a paper. There is suspicion that in some states, *nothing* is ever rejected. This may be unfair to some disciplines, but I think it is accurate for mathematics.

D. just could not bear the Banach–Tarski theorem: that it is possible to take a sphere, divide it into a finite number of pieces, and then reassemble the pieces to form two spheres, each exactly as large as the original one. This seems paradoxical, and so it would be if we were talking about physical spheres made up of atoms. But when we have spheres made up of uncountably many points, each with no extension, the paradox disappears. A line one inch long contains *exactly* as many points as a line two inches long. That does not bother me, it does not bother any mathematician, students of mathematics can be made to appreciate it, and its paradox content is the same as in the Banach–Tarski result. D. realized that the Banach–Tarski theorem was a theorem of set theory, so if he could upset the foundations of set theory, the Banach–Tarski theorem would have to go.

To get rid of it, D. chose to prove that the set of real numbers is countable, and that Cantor's diagonal process is a snare and a delusion. This would eliminate the Banach–Tarski theorem from mathematics, along with several other things. Stripped of its verbiage, D.'s proof is the same erroneous one that bright undergraduates sometimes make: D. argued (in effect; the verbiage makes it not all that easy to see) that if we list the real numbers between 0 and 1 as follows,

$$.1, .2, .3, .4, .5, .6, .7, .8, .9, .10, .11, .12, .13,$$
$$.14, \ldots, .99, .100, .101, \ldots, .999, .1000, \ldots,$$

putting down first the one-digit decimals, then the two-digit ones, and so on, we will have a list of all the real numbers between 0 and 1. The reply to this argument—which usually elicits an "Oh" after a few seconds' thought from bright undergraduates—that the list contains only the terminating decimals and none of the non-terminating ones, might not affect D. at all. His article reads as if it is by someone convinced, whose mind is not going to be changed by anything. It is by, in two words, a crank, and it is no credit to the state of X.

CONGRESSIONAL RECORD, MATHEMATICS IN THE

Members of Congress do not offend their constituents if they can help it, which explains why this appeared in the *Congressional Record* in 1940:

> Mr. [H.]. Mr. Speaker, one of my good friends, Mr. [M. B.], of [Wyoming], has recently discovered a new relationship between pi and *e* and in order to date, as well as to give a partial but necessarily incomplete description of his discovery, I ask unanimous consent to insert the attached statement by Mr. [B.] in the *Record*.
>
> Mr. [B.] claims that this is the first basic mathematical principle ever developed in the United States.

There follows a whole page of fine print. The title is

DIRECT RELATIONSHIP BETWEEN THE CIRCULAR
AND HYPERBOLIC MEASURES

and it begins

> In using mathematics, a number of relations of different numbers are known to exist, while other relations are felt to exist but have so far eluded exact definition.
>
> Euler (1797–83), who evolved the figures now used for cannon fire, indicates the strange interrelation between the number 3.1416, etc., which is the result of dividing the circumference of a circle by the diameter, and the number 2.178 (whose hyperbolic logarithm is 1). Euler uses the square root of minus 1 to arrive at his relation and therefore arrives at no definite conclusion. The square root of minus 1 is the base of many figures of

mathematics but defies definition. It is used, however, to correlate phenomena that are known to exist and may be said to be the base of Poncelet (1788–1867) reciprocity and Riemann (1826–66) surfaces.

B. gave relations between π and e not involving i and hence more useful than $e^{\pi i} = -1$ since, as he said, i defies definition. "2.178" is a misprint for "2.718," and Euler's year of birth was actually 1707.

Here are two of B.'s relations between π and e:

The square root of 5.4321 is 2.33 with a remainder of .32 (decimal points disregarded), which are the logarithm of 10 to the base 2.718 and the reciprocal of 3.14159 (the circumference of circle divided by the diameter), respectively.

The square root of 9.8765 is 3.142, and the reciprocal of the remainder is 2.3, with 2.72 as a remainder, which figures are the circumference of a circle divided by its diameter plus the reciprocal of the logarithm of 10 to the base 2.718, with a remainder of 2.72 which is the hyperbolic base or number whose hyperbolic logarithm is 1.

Furthermore, the sum of those two numbers is 15.3086—one-tenth of which is 1.53, which is as high a value as the transcendent angle may reliably assume.

The first paragraph seems to say that

$$\sqrt{5.4321} = \ln 10 + \frac{1}{10\pi}.$$

This gives

$$\pi = \frac{1}{10\left(\sqrt{5.4321} - \ln 10\right)} = 3.5585\ldots,$$

which is not even close. If I am reading it correctly, the second paragraph gives a value of π that is very, very far from $3.14159265\ldots$. I think that I must have misinterpreted B. The third paragraph says that

$$\frac{5.4321 + 9.8765}{10} = \frac{\pi}{2},$$

likewise not very close.

B. continued:

Mathematical respectability and correlation of these remainders is given by the equation (McMahon, 1908) the derivative of X divided by the square root of the square of X plus the square of a is the sectorial measure, the derivative of whose hyperbolic sine is the ratio X to a.

When you differentiate

$$\frac{x}{\sqrt{x^2 + a^2}}$$

to get what B. calls the sectorial measure, you have

$$\frac{a^2}{(x^2 + a^2)^{3/2}},$$

and when you differentiate the hyperbolic sine of that, you get

$$\frac{-3xa^2}{(x^2 + a^2)^{5/2}} \cosh\left(\frac{a^2}{(x^2 + a^2)^{3/2}}\right),$$

which does not look anything like x/a. What did MacMahon really say, and how did B. get it wrong?

B. seemed to be one of those people whose heads contain undigested mathematics:

The above orthodox (partially) Riemann-Cauchy surfaces necessitate the use of two or three foci which are indicated by taking unity and dividing into 10 units according to the gudermanian $10(N + 1)$ with 5 on each side of the center to obtain 5.4321 and 9.8765. Though this phase of the problem is unsolved the answer may lie in temperature phenomena unavailable to me.

Where did B. run across the gudermannian? Many mathematicians could not define a gudermannian without looking it up, and some would not even know where to look. However, even looking it up would not help, since N appears nowhere else in B's document. Its value is thus a mystery and the gudermannian of a mystery is also a mystery.

You may be thinking that there is no harm in B.'s appearance in the *Congressional Record* because no one actually *reads* it. (That may not be strictly true: there may be some poor people who must read the *Congressional Record* as part of their job.) Filling up a page of the *Record* probably made B. feel good, kept a vote or two for Representative H., and cost the taxpayers only a few dollars for typesetting, paper, and printing—insignificant when divided among the taxpaying population. Ah, but the one appearance in the *Record* was not all. Any citizen may, on request and by paying the expense involved, get reprints of part of the *Record*, and that is what B. did. I know that he did that because my copy of his statement is headed

(Not printed at Government expense)

and it was sent to the mathematics department at my school. No doubt many copies went to people who were as ignorant as Representative H. of the quality of the contents, and the official look of a document headed "Congressional Record" in large, boldface Gothic type might fool them into thinking that B.'s vaporings had governmental approval, or even that they had the force of law. Surely Representative H. or one of his assistants had a form letter, skillfully designed to turn down without giving offense, that denied people access to the pages of the *Record*. It should have been used.

I am fortunate enough to have what must be the last three pieces of B.'s correspondence still extant. His letterhead had on it his name, address, and specialties:

Mine Roof
Mechanics
Mining System

Representative H.'s insertion in the *Record* had done its work:

Inquiries about [H.]'s remarks (pi & *e*) necessitate a pamphlet which is going to press. Correlating Heat, Light, Sound, Electricity, Atomic Structure and Gravity. Three states of matter co-existent on surface of solid explained as byproduct. Igneous Geology put on sound basis and exposing two mysteries of astronomy, moreover sun to earth problem solved conclusively.

That was sent to a mathematician who, correctly, did not reply, provoking B.'s response:

The enclosed card was sent you some time ago. I wish to make sure that you are aware of this pamphlet and believe it no more than courtesy that you acknowledge receipt though uninterested.

B. was clearly a bluff, no-nonsense person (when away from mathematics), no doubt well-suited for the life of a mining engineer in the Wyoming of a half-century ago.

CONSTANT SOCIETY, THE

There are people who think that the accumulated wisdom of the human race is summed up, in its entirety, in its proverbs. Here is evidence for that position, the proverbs being "All things come to those who wait" and "There is less here than meets the eye."

In 1978 there was a full-page advertisement in the *American Mathematical Monthly*, half of it being taken up with a huge π, the rest with text, which included

$$\ldots \pi, \ e^x, \ i, \ 0\,k, \ c, \ G, \ldots$$

Constant Processes

Being absolutely important yet deceptively assumed, Constant Processes advocates the exclusive study of the fundamental constants. The book is a collection of expressions of theorems and formulae exhibiting the remarkable universality of the constants.

And so on, ending

Also included, is the new and revolutionary result: proof that e is the optimum number system base.

The book, by T. D., was available from The Constant Society, whose address was a post office box, at a terrific price for 1978: $29.50. It was too high for me, even to find out what $0\,k$ was or how to represent numbers in base e, so my curiosity went unsatisfied and the Constant Society slid out of my consciousness.

Until, unexpectedly and much later, information came from a mathematician who knew D.:

[D.], I'm sorry to say, is a local boy. He is a man of boundless enthusiasm for mathematics and very little talent for, or understanding of, the subject. He spent about ten years as an undergraduate here at the U. of [W.], finally managing to cobble together a bachelor's degree in mathematics. I still see him around occasionally; I think he's working as a computer programmer somewhere. At some point he got fascinated by π and e and the many strange and lovely formulas involving them—which is all very well, but he got stuck in a rut and started taking a decidedly mystical interest in these numbers and formulas. The upshot was that he wrote a book consisting of nothing but formulas, equations, and occasional statements of theorems, scrawled out in his wretched handwriting. Some of the formulas—the ones he copied out of integral tables or other sources—are correct; some of them—the ones he worked out himself—are not; none of them are accompanied by any proofs or references. So most of the book makes for pretty dismal reading.

For a sample, here is part of page 1.37:

$$\frac{\pi}{\sqrt{2}} = \int_{-\infty}^{\infty} \frac{dx}{1 + x^4}$$

$$\frac{\pi^2}{4} = \int_0^1 \ln\left(\frac{1 + x}{1 - x}\right) \frac{dx}{x}$$

$$\frac{\pi}{4} = \int_0^{\infty} \int_0^{\infty} \sin(x^2 + y^2)\, dx\, dy$$

$$\frac{\pi}{4} = \int_0^{\infty} e^{-x} \frac{\sin x}{x}\, dx$$

$$\pi = \frac{2^R (R!)^2}{(2R)!} \int_0^{\infty} \sin^{2R} \theta\, d\theta$$

Exactly what proverbial wisdom tells us: there is less here than meets the eye.

Here is why e is the optimum base of the number system, as explained by the mathematician who knew him:

I think I can explain it more clearly than [D.] does, so here goes.

Consider an old-fashioned adding machine, on which you enter numbers by punching numbers from a rectangular array: there's the ones column, the tens column, and so forth, and each column has buttons for the digits from 0 to 9. Say you want to enter any (whole) number less than a million: then you'll need 6 columns of 10 buttons, for a total of 60 buttons. But now envision a machine that works in base 2 rather than base 10. If you still want to go up to a million, which is a little less than 2^{20} you'll need 20 columns of 2 buttons, for a total of 40 buttons. Now 40 is less than 60, so base 2 is more efficient than base 10 in this respect. What is the *most* efficient base? Well, if you want to express numbers up to N in base b you need

$$b \log_b N = \frac{b \log N}{\log b}$$

buttons. Set the derivative with respect to b equal to zero, and you find that (no matter what N is) the minimum occurs at—yes, e! [D.] was firmly convinced that this result had the potential to revolutionize the computer industry.

All things come, if you wait long enough.

By the way, in the advertisement's list of constants, c is the speed of light and G is the constant of gravitation. The constant $0\,k$ may be more obscure: it stands for zero degrees Kelvin.

CONSULTATION, LACK OF, OF CRANKS WITH EXPERTS

Every file of crank mathematics in the country has in it a copy of a two-page paper with the shortest of all possible titles,

1

by J. B. He sent it to almost every mathematics department in the country, at great cost in paper and postage, not to mention wear and tear on his tongue licking stamps and sealing envelopes. Such mass mailings are not common. Most cranks hit only the high spots: Harvard, Yale, and the other places that come to mind when people think of big-time universities. But B. must have thought that his discovery was so amazing that it deserved the widest possible circulation. Here it is:

$$\sum_{N^E}^{(N-1)^E} \left(\sqrt[E]{N^E} - \sqrt[E]{N^E - 1} \right) = 1.$$

He went on to explain:

Following is the expansion of the above formula in descending order.

Example: $N = 3$, $E = 2$.

$$\left(\sqrt{9} - \sqrt{8} \right) + \left(\sqrt{8} - \sqrt{7} \right) + \left(\sqrt{7} - \sqrt{6} \right) + \left(\sqrt{6} - \sqrt{5} \right) + \left(\sqrt{5} - \sqrt{4} \right)$$

$$= (3.0000 - 2.8284) + (2.8284 - 2.6457) + (2.6457 - 2.4494)$$

$$+ (2.4494 - 2.2360) + (2.2360 - 2.0000)$$

$$= .1716 + .1827 + .1963 + .2134 + .2360 = 1$$

The number of iterations required for any problem is equal to the solution of $N^E - (N - 1)^E$.

The number of decimal places carried out are optional, but must be consistent for any given problem.

B. didn't notice that his sum was telescoping and that all of its terms cancel each other out except the ones at the start and finish. He didn't notice that all he had discovered was that

$$\sqrt[E]{N^E} - \sqrt[E]{(N - 1)^E} = N - (N - 1) = 1.$$

He didn't ask anybody about it at the local high school or college. He didn't send it to one or two places first. He didn't *consult*.

CRANK, CASE STUDY OF A

Let us arrange mathematical cranks from left to right. At the far right would be people like a mathematics teacher who absolutely insists that all his pupils master Descartes' Rule of Signs: it is, he thinks, of vital importance, and no one can claim to be educated in mathematics who is ignorant of it. This behavior is so far to the right that it hardly deserves the label of crankery: "crotchety" or "slightly eccentric" describes it more accurately. At the far left are people who are convinced that they have *the* truth, that it is revolutionary, that mathematicians are engaged in a vast conspiracy to suppress it, and that fame and wealth are rightfully theirs and that one day they will have them. Again, "crank" is not as descriptive as another word— "lunatic" in this case.

A graph of number of cranks versus position on the spectrum looks, I think, like a portion of an exponential curve e^{kx} (Figure 4). The median and mean cranks are found to the right of center. The precise value of k has not yet been determined.

The following is a description of G. B., a crank near the median or slightly to its left, by a professor of mathematics who knew him:

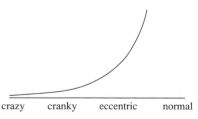

FIGURE 4

crazy   cranky   eccentric   normal

Mr. [B.] is a thin, active man of about fifty-five, a life-long bachelor who gets his living as some sort of engineer. If I recall correctly, he got his degree, B. S. I presume, from [an engineering school] after World War II, on the G. I. Bill, a consideration for which, among other things, he is lastingly grateful to the armed forces. He is an active member of a veteran's post. He smokes unfiltered cigarettes heavily, with no apparent ill effects, and gets by, he says, on two to four hours of sleep a night. He does not behave feverishly, and expresses himself with a certain, not inordinate, amount of care, in a style that may be admired in engineering circles.

Last fall, in October or November, Mr. [B.] announced a press conference and lecture, to be held on a Saturday afternoon in an American Legion hall, for the purpose of disclosing a new mathematical discovery. The event was announced in the newspapers, by an advertisement paid for by Mr. [B.], and on the bulletin boards of the mathematics departments of institutions around [his city], by a flier made by Mr. [B.] and sent to these departments with instructions to post.

His flyer was headed

ANNOUNCEMENT

A new mathematical discovery has been made in Number Theory

A SIMPLE METHOD OF DETERMINING ALL OF THE

	1)	PRIME NUMBERS
	2)	SQUARE ROOTS OF INTEGERS
&	3)	RATIONAL PYTHAGOREAN TRIPLETS

All from the numerical sequence:

$$1, x^2 + (-1)^p, \ldots x a_{n-1} + (-1)^t a_{n-2}, \ldots$$

Invited were "All Press, Radio, TV, & general news media" as well as "all Mathematicians, Public Officials, & other interested persons." The program was

12:00 noon–1:00 pm   News Media set-up

1:00 pm   Registration (no charge)

2:00 pm   "PRIME NUMBERS"
poster illustrated lecture
by the author

3:00 pm   Discussion—questions and answers

The professor continued:

For no reason other than to occupy the early part of the afternoon
before a departmental picnic that day, I went to the announced press
conference and lecture, and made Mr. [B.]'s acquaintance. As it happened,
I was the only person other than Mr. [B.] to show up, except for a friend of
his, a retired colonel, who was there when I arrived but soon left, saying
"I don't understand any of it, [G.]". Mr. [B.] and I talked for about three
and a half hours, counting an hour's interruption during which Mr. [B.]
delivered his lecture to me.

During our talk Mr. [B.] told me a number of things which were
verifiably true, such as that 39, 760, 761 is a Pythagorean triple, and that
the difference between the "hypotenuse" and either of the "legs" of a
primitive Pythagorean triple is either the square of an odd integer, or half
the square of an even integer.

Since in a primitive Pythagorean triangle (Figure 5), that is, one in which x, y, and
z have no common factor,

$$x = 2mn, \qquad y = m^2 - n^2, \qquad \text{and} \qquad z = m^2 + n^2,$$

where m and n are integers not both odd or both even, B. was rediscovering

$$z - x = (m - n)^2 \qquad \text{and} \qquad z - y = 2n^2 = \frac{(2n)^2}{2}.$$

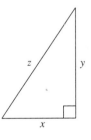

FIGURE 5

The professor went on:

(Our only falling out occurred over this last fact. After Mr. [B.] had made
the assertion, I saw how to prove it, and told him so. He said, "You
won't find that in any math book!" rather decisively, and wouldn't let
me show him how to prove the assertion. He had discovered the fact,

and had apparently convinced himself of its truth in his own inductive, experimental way, and wanted to consider the discovery his own.) Mr. [B.] told me that he had given no thought to mathematics from his college days up to one night five or six years ago, when, having coffee with his brother at 3 o'clock in the morning in an all night restaurant, after a double date during which much beer was consumed, he turned to his brother and said something like, "[F.], the Pythagorean triples can be divided into classes." He meant not just any old classes. Something had evidently dawned, or erupted in his brain, and he spent the next three years at hard labor, trying to find out what it was. During this time, and afterward, he presented himself and his partially completed work to mathematicians at [schools in his city] and possibly elsewhere. A few of these mathematicians were extremely impatient and discouraging (although [B.] is indefatigable and not easily discouraged, owing partly to his constitution and partly to an evidently very supportive circle of friends), but others at least listened, checked some of his claims, and directed him to certain books. He began to spend, and still spends, a good deal of time in mathematics libraries, poring over number theory books. This study has lately, since that meeting last fall, directed his attention away from Pythagorean triples and square roots, and towards prime numbers. The time he has spent with books has contributed nothing to whatever powers of insight he possesses, and has shown him that many of the things he took for his own, original discoveries are well known, in various forms. I suspect that he understands what he finds in the number theory books only when he recognizes his own work.

If I recall correctly, Mr. [B.] said that his first success, coming two or three years after that night when the virus first took hold, was the devising of a method for approximating square roots (rational approximations of square roots of integers; his method holds up for many numbers, but gives rational approximations only of rational numbers). In this matter of square roots, Mr. [B.] has done good work, phenomenal work considering the tools and methods at his disposal, but he has arrived on the scene two or three centuries too late to gather the applause of a grateful nation.

What B. had done with square roots was to rediscover how to get better and better rational approximations of them by using recursion relations to generate successive terms. For example, he approximated $\sqrt{3}$ with fractions a_n/b_n, starting with $1/1$ and $2/1$ and then using

$$a_{n+1} = 4a_n - a_{n-1}$$

$$b_{n+1} = 4b_n - b_{n-1}$$

to get the sequence

$$\frac{1}{1}, \frac{2}{1}, \frac{7}{4}, \frac{26}{15}, \frac{97}{56}, \frac{362}{209}, \frac{1351}{780}, \frac{5042}{2911}, \frac{18817}{10864}, \frac{70226}{40545}, \ldots$$

It converges very quickly, which allowed B. to say:

The famous Archimedean ratio of $\frac{70226}{40545} \approx \sqrt{3}$ (error $= 2 \times 10^{-10}$) is confirmed by Eqn 1.0. 40545 is the 9th term of the Square Root Sequence $D_{1,4,0,1,n}$ and the 8th term is 10864. Better yet, though, from the 11th and 12th terms of the Square Root Sequence, we have

$$\sqrt{3} \approx 1 + \frac{1542841}{2107560}$$

(error $= -6 \times 10^{-14}$).

The professor continued:

Mr. [B.] told me that if he laid out this method for generating Pythagorean triangles as thoroughly as he laid out the procedure for finding square roots, the exposition would run to fifty pages at least. From my conversation with Mr. [B.], what he does is something like this. Taking some sort of cue from some sequence of numbers, gotten by plugging certain initial parameters into some formula determining the sequence, he generates an infinite sequence of Pythagorean triples. Luckily, you only have to know the first few members of this sequence to carry out, or at least start carrying out, the next step. In the next step, he generates a second infinite sequence of Pythagorean triples, and he knows how to do this from knowing what the first sequence of Pythagorean triples was, and from some stuff contained in one or two auxiliary sequences over on the side. He continues to generate infinite sequence after infinite sequence of Pythagorean triples. He claims that each of these sequences of triples contains the preceding as a subsequence. (Obviously, that can be arranged.) All this sounds pretty uninteresting to anyone outside the any-information-at-all mathematical scavengers, who might be interested in a

theorem reading, if you do the following things, you get a bunch of Pythagorean triples; but Mr. [B.] did make one interesting claim with regard
to his Pythagorean triple machine. Take one of those infinite sequences
of triples produced at some stage of the process and, for each triple in the
sequence, form the right triangle corresponding to it. [B.]'s claim (I don't
remember it exactly) is that as you go down the list of triples, one of the
angles in the corresponding triangles, either the smaller or the larger of
the two acute angles, monotonely increases.

This is not as interesting as it might appear, since B.'s recursion relations give sequences of triangles converging in shape to a limiting triangle, and thus the sequence
of angles converges also. It is not very surprising that the convergence should be
monotone.

Besides that attractive claim, there is something else to recommend
a study of [B.]'s awful triple-generating process, and that is Mr. [B.]'s
assurance that it works. On this matter, he is as sure and loquacious as he
is on his process for finding square roots. Putting aside my earlier remark
about mathematical scavengers, his assurance makes me think that it may
be that the effectiveness of the bloody algorithm rests on an interesting
mathematical foundation. [B.]'s method for doing square roots, when
translated into familiar terms, is not uninteresting, even if the ideas under
it are well known, and the same may be the case with this Pythagorean
triple machine. However, I myself was not interested enough to try to
extract the whole story from Mr. [B.].

When it comes to prime numbers, Mr. B. is much less sure. He
attempted to devise a process, similar to the square root finding and Pythagorean triple generating processes in that it employs, in arcane ways,
sequences defined by formulas involving parameters into which values are
plugged at various stages of the process, for infallibly producing prime
numbers. The procedure he has come up with has produced two composite
numbers so far. He shrugs this off.

One thing that B. came up with was the sequence

$$a_n = 3a_{n-2} - a_{n-4}.$$

With $a_1 = 1$, $a_2 = 0$, $a_3 = 4$, and $a_4 = 0$, you get terms

n	1	2	3	4	5	6	7	8	9	10	11	12	13	14	15
a_n	1	0	4	0	11	0	29	0	76	0	199	0	521	0	1364

These have the property, as far as the table goes, that a_n divided by n leaves a remainder of 1 when, and only when, n is prime. $\frac{4}{3}$, $\frac{11}{5}$, $\frac{29}{7}$: remainders 1; $\frac{76}{9}$: remainder 4; $\frac{199}{11}$, $\frac{521}{13}$: remainders 1; $\frac{1364}{15}$: remainder 14. Have we found a prime-producing formula? Is it true that n is prime if and only if

$$a_n \equiv 1 (\bmod n)?$$

Unfortunately the answer is no, but the first exception does not arrive until $n = 705$, and the one after that is at $n = 2465$.

Mr. [B.] is not of that class of people who strive for easy fame by producing quick proofs of Fermat's Last Theorem or other famous conjectures. His non-membership in that sordid club is due not alone to his ignorance of famous conjectures, but also to a measure of rigor in his make-up. He works only on matters to which his thoughts lead him, and he works hard to produce true somethings. I say "somethings" here because what he produces are not propositions, although his work can be translated into a pile of propositions. In the course of my first talk with him I mentioned Fermat's conjecture, and the twin primes problem. He thought briefly and said that those were not the kinds of things on which he would have the slightest idea how to proceed. In our two or three talks since then, he has never mentioned these famous conjectures, although he has mentioned one other, Hilbert's Tenth Problem. I'll get to that in a bit.

Mr. [B.] does expect to profit from his endeavors. His idea is to sell his discoveries. After our last session, he said that he expected to find "the secret of the primes". I asked him what he would do when he found it, and he replied, "Sell it." I asked who would buy it, and he said, "Wouldn't you pay to learn the secret of the primes?" I think his plan is to sell copies of a treatise, not to sell the secret in one lump, although he didn't make that clear. Anyway, the important thing about his motives is that his mind is not spoiled by greed. He is constrained by a clear notion of truth. If we are to judge him a mathematical crackpot, we should make sure to specify that he is so judged not because his mind is unhinged and he happens to discharge the wreckage of a shattered consciousness in musings mathematical, but because he is a hard laborer, and overly enthusiastic, who happens to be totally ignorant of the mathematical discipline relevant to his personal interests. He can communicate with mathematicians only through considerable effort on their part. Thus he and his work, which is not without merit by any means, are regrettably, but decisively, isolated from the only fraction of the population that might be interested. Given his age and habits of mind, it is unlikely that he will ever assimilate the minimal bag of tricks and learn the minimal amount of lingo necessary for him to converse easily with mathematicians and apply his talents within the practical boundaries of contemporary mathematics.

Mr. [B.] apparently thinks, or thought at the outset of his endeavors, that mathematicians spend their time doing square roots, computing logarithms, and things of that nature, and that the discovery of a new way of doing square roots or computing logarithms constitutes a major mathematical discovery. In one of our sessions in my office, Mr. [B.] wanted to check a computation, and was quite surprised to learn that I had no logarithm tables, or lists of primes, or any of that stuff there in the office. His idea of what mathematics is manifests itself [in his work]. He intended that his methods be *used* to calculate square roots, produce tables of Pythagorean triples, and list the primes. He genuinely expected some sort of ovation from the mathematical community, which he envisioned squirming in its chairs for want of good methods of calculating square roots, tabling Pythagorean triples, and listing the primes. As he has now undergone some sort of initiation in the mathematics libraries in the area, his extreme naiveté has been tempered. He no longer, to judge from our last two meetings, searches for "useful" processes and algorithms; rather he now regards the positive integers as a natural system, of which he seeks to be the philosopher and expositor. In particular, he finds it marvelous that the primes don't occur in any regular way in the list of positive integers, and has set himself to discovering some order in their occurrence. What he has been doing lately is this: he considers the one-sided infinite matrix whose (i, j) entry is $i^2 + j$, and looks at the numbers that lie on lines, parabolas, and hyperbolas drawn on this array, with the idea of discovering patterns. He has discovered some God-awful patterns so far. I remember one of the conjectures I managed to formulate after penetrating, or so I hope, [B.]'s exposition of one of these patterns: for any positive integer m, there is a prime p such that

$$p + 1 + 1^2, \; p + 2 + 2^2, \ldots, \; p + m + m^2$$

are all primes. This is a much stronger conjecture than the twin prime conjecture, and is very likely false, although for no reason easy enough for me to make out. But I am no number theorist.

A portion of B.'s matrix is

2	3	4	5	6	7	8	9	10	11
5	6	7	8	9	10	11	12	13	14
10	11	12	13	14	15	16	17	18	19
17	18	19	20	21	22	23	24	25	26
26	27	28	29	30	31	32	33	34	35
37	38	39	40	41	42	43	44	45	46
50	51	52	53	54	55	56	57	58	59
65	66	67	68	69	70	71	72	73	74
82	83	84	85	86	87	88	89	90	91
101	102	103	104	105	106	107	108	109	110

and it is easy to see how he could have become fascinated by it. Look at that diagonal going up and to the right from 101—all primes! Look at that portion of a diagonal going in the same direction from 107 to 59—all primes again. The matrix cries out for extension to see if the whole diagonal contains nothing but primes. If it doesn't, then the matrix can be extended farther to look for another diagonal like that; and if one doesn't turn up, other straight lines, parabolic paths, or hyperbolic paths can be looked at. Of course, the numbers on any straight-line path through the matrix, or on any parabola or hyperbola, will be values of a polynomial, and we know, as B. did not, that no polynomial can have values that are all prime.

His latest work and his reading have brought Mr. [B.] to the recent report in *Scientific American* on Hilbert's Tenth Problem. The idea of a Diophantine equation whose solutions are primes, and of which all primes are solutions (excuse me, I'm a little shaky on Hilbert's Tenth Problem) has stirred some associations in his mind. He suspects that he may have hit upon some of the essentials of the matter, whatever they are, in his own work.

How Mr. [B.] came upon his present enterprise, finding patterns in the factorizations of numbers in that infinite matrix, I have no idea. I'm afraid it looks like a bottomless pit. He is striving for some description of the relationship between the numbers of that array that will indirectly describe how the primes fall. He can test the dependability of his patterns only by experiment for he has neither the tools for, nor the inclination towards, general demonstration. He can only describe what he has found in the rawest form, by showing you the array, for he has neither the tools for, nor the inclination toward, the elegant simplicity of expression that is the stuffing of mathematical work. He will be searching for patterns until Judgement Day, when he will, I hope, finally receive some suitable reward for his labors.

B. kept at it. Two years after the date of that letter there appeared a paperbound book, $8\frac{1}{2}''$ by $11''$, xvi + 48 pages,

ALL THE
PRIMITIVE PYTHAGOREAN
TRIPLETS

[B.]'S GENERATION METHOD (BGM)
BY PROGRESSIONS

The
complete
and final
solution to the
Pythagorean
mystery:

the systematic generation of all the
primitive Pythagorean triplets
in ordered families of
ordered progressions
using a unique
indirect but
absolutely
orderly
method

published by "B. G. [M.] Syndicate" in a numbered first edition of which my copy is #110, and dedicated to

The spirit of patriotism
the miracles of useful intellectual development
the wonders of intellectual development
individual achievement
without interference or subsidy

Of his three topics—square roots, Pythagorean triples, and primes—B. seemed to have given up on primes and put square roots aside for a time:

It is hoped that some day this book can be reprinted in a more professional manner, i.e., typeset and cloth bound, and that a related sequel by the author entitled "ALL THE SQUARE ROOTS BY SIMPLE DIVISION" may emerge.

He presented his work with humility:

> If anyone had told me twenty years ago that I would make unaided a new mathematical discovery even of a simple nature, such as this one is, and then write a book about it, I would have been speechless with disbelief, for I was under the prevalent impression that if it hadn't as yet been invented, devised, or discovered, it never would be.
>
> Nevertheless, one day out of a clear blue sky a power greater than me tapped me on the shoulder, so to speak, and started me off with the absolute and overwhelming belief that there existed what I refer to as a "complete and final solution to the Pythagorean mystery".
>
> The pleasure of working tens of thousands of hours on it was extreme, since I already knew with absolute certainty that I would formulate the detailed solution.
>
> I do hope this book is readable and understandable and that the format and style are acceptable. Whatever errors, typographical or otherwise, you might encounter are my fault, since this entire undertaking (the reproduction and binding excepted) is the product of one sole person—me.
>
> My fervent hope, however, is that this book which you have elected to read and at this moment hold in your hand is useful to the extent that perhaps some little item, or even a facet thereof, contained herein will generate in a mathematical mind a tiny spark which will eventually lead to a really worthwhile mathematical discovery.

How one wishes one could take G. B. aside and tell him, "[G.], this book has some really good things in it. It's going to be useful to a lot of people, and you're right—maybe something big will come from it some day." How one wishes one could give B. some repayment for his tens of thousands of hours of work (tens of thousands of hours!—a year's work, full time, amounts to two thousand hours), selfless work, meant only to advance mathematical science. How one wishes B. could be told that his time, a significant part of his life, was not spent in vain.

But one cannot tell him that. His book is filled with pages of numerical calculations based on his equation

$$D_{c,x,a,p,m,n} = xD_{c,x,a,p,m,n-2} + (-1)^p mD_{c,x,a,p,m,n-4},$$

giving pages and pages of Pythagorean triples, arranged into families. The equation certainly works—the front cover of his book is decorated with a right triangle with sides

235887607205063197716l920,

3145168096069599348102721,

and hypotenuse

393146012008605866477932g.

But no one told B., or if he was told he did not hear properly, that any Pythagorean triple (x, y, z) comes from

$$x = 2mn, \qquad y = m^2 - n^2, \qquad \text{and} \qquad z = m^2 + n^2,$$

so there is no trouble in generating them, and there is no more interest in arranging them in families than there is in arranging perfect squares in families according to how many 7s they have in them. As far as I know, there was no second edition of the book.

CRANK, THE MAKING OF A

One explanation of the phenomenon that some people are mathematical cranks is that they were born so, as people with cleft palates or club feet. On the other hand, it is possible that some are made and not born. As to how they are made or who makes them, consider the following example.

W. S. proved the four-color theorem, he thought, and published his proof in a 16-page booklet, copyrighted in 1982, and available from the author for $6 a copy, later increased to $10. The preface to the booklet gives the history of his efforts.

> As a person who terminated his formal education at mid-high-school level back in the 30's (the Great Depression days), I am proud to be able to present this short proof of the four-color conjecture.

> My work on the 4CC has no doubt been an ego trip all the way—a reflection of my disappointment at having to leave school prematurely, when my heart had been set on a career in mathematics and physics.

> It may have been the ideal career for me: at fourteen I had discovered and was using calculus before I could even name what I was doing—and I had worked out a series for calculating logs by studying a slide rule. (Don't ask me what that series was!)

> It was this stubborn primitive egotistical desire to show what one can do on one's own, before or without consulting the works of the masters, that resulted in my proof of the 4CC.

So far, you may think, we have the typical crank. Formal education terminated at an early age, no training at all in mathematics beyond a little algebra and geometry, getting on in years, attracted by a difficult problem and deluding himself that he had solved it: all signs of crankhood. However, read on:

My fascination for this problem developed only after I had learned in September, 1976, that Professors Kenneth Appel and Wolfgang Haken's computer proof of the 4CC had been presented at the University of [T.].

It was a marvelous feat, assisted by more than three quarters of a century of contributed input. But my obstinate refusal to believe that this conjecture required anything more than a short simple proof, obsessed me to spend all my spare time working on the problem.

It is *not* one of the marks of a crank to acknowledge that anyone else's work on the crank's problem has any value at all. Also, thinking that a shorter proof of a theorem ought to exist and spending time trying to find it is not cranky at all. Hundreds of mathematicians do the same thing.

S. then went to inquire at the mathematics department at a nearby research university. This is also something that cranks do not do: they do not *ask* mathematics departments for information, they *give* it to them. A member of the department

presented me with some literature on the history of the problem, which, as indicated above, I delayed reading until I was sure it would not interfere with my own progress.

I quickly developed a Hamiltonian circuit approach (not knowing at this time that this was what it was called) where I sought to spiral a single two-color chain around. . . .

Prof. [B.] kindly tolerated my flood of correspondence attempting to show that I had a proof, until March, 1978, when he recommended that since the four-color problem was not in his field, I should send copies of my work to Prof. [T.], at the University of [W.]

Professor B. did not suddenly discover that the four-color problem was not in his field. He knew it all along; so why did he tolerate a flood of correspondence for eighteen months? Perhaps he tired of S. and chose to get rid of him in a way that would cause minimum inconvenience.

Prof. [T.] correctly advised me that I had *not* written "a systematic and logically tight account", and would not be "taken seriously" until I had done so.

Here again is a non-crank characteristic. Cranks *never* agree that their works are not flawless, and *always* blame any failure to understand them on others.

Another thing that cranks tend not to do is try to have their efforts published through the usual channels. Not so with S.:

My proof of the basic conjecture of this approach was still far from "obvious" when I first overzealously attempted in the summer of 1979

> to have my first paper containing it published in [a mathematics journal],
> mainly on the premise that I could not find a map in which it could not be
> shown to be true, including a large configuration contained in the above-
> mentioned Moore map. . . .
>
> Again I was rightly upbraided by Prof. [T.], who refereed my paper,
> for substituting the word "obvious" for a rigorous proof.

That sounds as if S. had found a method for four-coloring maps, one that worked
for any map he tried it on. This is of course not the same thing as a proof of the
four-color theorem. Note that S. was still agreeing that what the authorities were
telling him was correct.

Now S. started his descent into crankhood.

> My next exploit was merely to expand this paper somewhat to show
> how it applied to a random map of my own design, and send out over one
> hundred copies to highly rated math departments in institutions through-
> out the U.S.A. and Canada.
>
> Two acknowledgements only: I thank Henri Petard of Princeton Uni-
> versity for his friendly reply.

The author of that friendly reply was guilty of a crime that ranks somewhere between
lack of compassion and cruelty. There is no Henri Petard at Princeton. "H. Pétard"
is the pseudonym of the author of "The mathematics of big game hunting," the still-
remembered classic of mathematical humor first published in 1939. What I think
happened was that S.'s work fell into the hands of a graduate student at Princeton
who decided to have a little fun with him. I imagine that the reply was, in the
graduate student's mind, a masterpiece of irony, one that he delighted in showing to
his fellows. Poor S.

> My greatest appreciation goes to Prof. [C.] of the University of [A.],
> without whose encouragingly analytical reply I might not have continued
> my efforts. I sent him my first rough draft of my first version of a complete
> proof, on January 1st, 1980.

That was S.'s last mention of Professor C. What was Professor C. encouraging?
The descent continued:

> Again copies went out to math departments: Henri Petard again ac-
> knowledged receipt, but this time I also received acknowledgement from
> [H. S.], chairman of the department at [S.] University, and from the de-
> partmental secretary, [M. M.], at [H.] University. . . . Thanks.

The poignance is hard to bear: think of S., every day eagerly waiting for the day's
mail, riffling through the letters searching for one from a mathematics department,

every day disappointed, every night hoping that one will come the next day. Then one day there arrives a letter with H. University as its return address! He opens it, his heart pounding; he unfolds it, his hands trembling slightly; he reads it: it is a form letter that the departmental secretary has been instructed to send out in response to unsolicited manuscripts. He is grateful. He gives thanks. One's heart bleeds.

> But the most helpful and encouraging reply this time was from Prof. [K.] of the University of [C.], who complained mainly about my lack of lucidity.

There! The first sign of the bitterness that invades and poisons the minds of cranks. It spread:

> So I churned out another paper that I thought would take care of that point—it didn't!
>
> I randomly distributed copies of this paper at the Universities of [T.] and [W.], in the summer of 1980. Result: One request from a student at [W.] that I send him a copy.
>
> I then mailed copies to nearly every member of the faculty at [W.] who was in Combinatorics. . . .
>
> I was convinced, when Prof. [S.] phoned me, that I had at last found someone who would give my paper a competent reading. Alas, Prof. [S.] merely wished, without ever thoroughly examining my paper, to convey to me that whatever my solution was, it could not possibly be correct, *since the statistical odds against anyone having an elegant solution were enormous.*
>
> I am really not very perceptive. It took nearly three months' correspondence for him to convince me that he really had no intention of ever giving my paper a competent reading.

First S. tried for publication in a journal, then he distributed his revised proof widely, then he sent the next revision to universities in his vicinity, and now the latest version went to high schools:

> So, back to the drawing board!
>
> Early in 1981 I turned out a version with a title that was intended to be somewhat tongue-in-cheek: "A Short "Proof" of the Four-Color Theorem for Primary and Secondary Schools". This was sent to five high school mathematics teachers, among the first.
>
> My thanks for replies from four of them, two of whom stated they could find no flaw in my argument (at the same time acknowledging their lack of qualification in the field). All four of them stated they thought it deserved a competent reading.

Next came typically cranky behavior:

> I ran through a whole gamut of gimmicks trying to get a compe-
> tent reading of this paper, from posting illustrations on university notice
> boards, to making bets as high as $1000.00 that my findings could not be
> disproven. . . . No takers.

The process might have petered out, but one person referred S. to a mathematician of wide renown, who in turn referred him to Professor B., another mathematician of eminence:

> The last week of June, 1981, I sent Dr. [B.] a copy of my proof.
> Early in January, this year (1982), I received from him the first strong
> indication that it was a success, when he wrote: "If your work were to
> contain significant results, I would be happy to sponsor it in my capacity
> as editor of [a research journal]."
>
> Never before had my proof been given an "if". Never before had
> there been anything but the unquestioned pre-assumption that surely it
> must be flawed—"significant results" of any kind had certainly not been
> considered possible.

Thus does a sequence of actions, individually harmless or even praiseworthy,
combine to produce unfortunate consequences. It is decent to reply to letters; it
is nice to suggest someone else who may be able to deal with a matter in which
your competence is limited; statements such as "If you have in fact proved the
Riemann Hypothesis, then your work should be published" are undeniably true.
S. was inflamed. His head must have been full of visions of recognition, fame, and
vindication, of fantasies of appreciation and renown, with the prospect of interviews
and—who knows?—appearances on national television.

In an unfortunate coincidence, the American Mathematical Society had sched-
uled a meeting in S.'s city. He prepared a ten-minute presentation and got Professor
B., as a member of the Society, to agree to introduce him. B. seems to have had
some misgivings:

> The notices and abstracts of the AMS indicate or suggest that I spon-
> sor your claim (at least) of a short four color proof. You have misunder-
> stood. My offer of sponsorship was much more limited—I thought it might
> be possible (but perhaps not likely?) to help you publish in a professional
> journal a modest result based on your method. This would do a great deal
> to establish credence for you, and moreover, an attempt on the four color
> conjecture could build on what was already done. Please change your
> tactics, or your chances of being taken seriously will fly out the window.

But it was too late. The ten-minute talk was given, and I think that it gave S. the final push into irreversible crankhood. About the talk he said:

> Preceding the presentation of the report indicated above, the session chairman read an announcement provided by Dr. [B.] that he (Dr. [B.]) actually *did not support my claim to a short four-color proof.* . . .
>
> The most significant point in [Dr. B.'s paragraph above] is Dr. [B.]'s submission that an attempt on the 4CC can *"build on" what I have done.* But he appears here to deny me credit for even this "modest result", simply because I refused his offer to publish, in favor of self-publishment, and managed to gain the privilege of presenting this paper, in spite of his attitude.
>
> It is obvious that I still firmly believe that what I have accomplished does not need to be "built on" to establish *final* proof of the 4CC.
>
> The attention given my presentation, and the appropriateness of the questions, adequately refute Dr. [B.]'s claim that I would not be taken seriously.

Giving the ten-minute talk was so tremendously exciting that S. became flustered.

> I was asked a question that I did not correctly interpret at the time, so I feel I must have answered it rather stupidly. The question as I now interpret it: "Could what I offer as a proof be a non-proof because subsequent steps in the method produce a circular negative result?"
>
> I apologize to the questioner (name not noted), but the answer is "no". I think the method of map reconstruction used in this presentation gives ample evidence that subsequent steps may *alter* but *not negate* the *positive* results of previous steps.
>
> *NOT ONE PERSON* VENTURED TO POINT OUT *A FLAW* IN MY PROOF.

>   . . .

> Dr. [B.] had been doodling during the presentation, after which he handed me two maps he believed could be further counterexamples.
>
> I must admit here to having become too mentally exhausted to follow my own logic. I failed to do in a few minutes what I could normally have done in a few seconds: show why they were not counterexamples.

B. had now become a villain and S. had now taken to writing as cranks sometimes do: in short paragraphs, with copious underlining, and hints of persecution and conspiracy.

Dr. [B.] has made such a show of attacking flaws merely in the writing and style of my paper where lack of professionalism is noted (presuming I am writing for a professional journal), that I can only accept these attacks as *red herrings* meant *to distract attention* from the fact that *he has never been able to find a single significant flaw in the proof itself.*

He has never pointed out such a flaw to me, unless it lies in his ostensible doubt that I have exhaustively analyzed all cases.

If this professed doubt arises from the fact that I have not *listed* and *enumerated* all cases, *only here* is there any reasonable justification for his attack on the writing and style of my paper. . . .

I have no doubt that there were many at my presentation who have just as great a "vested interest" in the computer proof as [Dr. B.]. It would be very hard for them to acknowledge the success of an amateur. But I sincerely hope that a "gentlemen's agreement" (in this case including the ladies) has *not* brought about *a conspiracy of silence* regarding my work. I should by now, more than three weeks after the presentation, be receiving very strong *negatives* regarding my proof, *if there were any to give.* I have received *none.*

There is no doubt in my mind that all who had any interest in the 4CC who had attended my presentation went home to their desks or computers to find *where I must have gone wrong.*

Are they still trying?

By now that should represent several years of man-hours! Longer than Dr. [B.] put in!

I am not a Velikovsky in a science where high probability is the criterion.

I am not a Sister Kenny battling a profession hardened against anyone who questions its authority. Or am I?

I have decategorized what the profession would like to continue to regard as a most complex mathematical problem—I have simplified it! *That is my sin.* I have destroyed a myth!

The problem is that *it is possible for less than brilliant minds to see that I have done this.*

So in the name of integrity, I trust it will not be long before I start receiving confirmations.

Indeed it will seem very strange, to others as well as myself, if persons of the high intellectual calibre I am dealing with should take an inordinately long time to recognize a simple proof, and place the *credit* for it *where it belongs.*

Here's to the honest person who first breaks from the pack with the truth I am waiting to hear!

Will it be *you* who is reading this now?

S. overestimated the amount of time mathematicians will spend on problems other than their own. I'm sure that no one went away from his ten-minute talk to a desk or computer to work on his proof. Indeed, why should anyone? Cranks misinterpret indifference as evidence of conspiracy, but it is only indifference. It is painful to be ignored, but cranks should not think that highly trained and busy professionals should drop everything to pay attention to someone with no training or qualifications in the field, for free.

S. then became a complete crank, sunk in a morass of rage, frustration, and delusion. Here is a form letter that he composed. I do not know how many copies of it were sent, if any:

Prof. X. Pert
Faculty of Mathematics
Whatever University, Whatever Place

Dear Prof. X. Pert,

Enclosed is a copy of my proof of the four-color theorem, presented at [the meeting of the American Mathematical Society], plus a rough draft of the supplementary insert that will accompany remaining copies of the first edition.

Enclosed also is a copy of the ad I intend to place in your university student newspaper, the Blabbit, as well as in student newspapers of universities with high mathematics ratings throughout the U.S. and Canada—unless I am given sound reason for not doing so, within a reasonably short time. [The text of the advertisement is

I WILL PAY
$10,000.00
to any math person able to find
a significant flaw in my short
proof of the four-color theorem
sufficient to invalidate it as a
PROOF
Must have purchased copy
$10.00 Money Order to:
[W. S.]]

The Blabbit may have already received this ad copy, together with the letter (copy) here enclosed containing my instructions to keep the ad on file till I give final confirmation that I wish to have it published. You may expect, accordingly, to be interviewed by this paper's reporters, regarding my offer of such a large award.

Enclosed also is a copy of the report I published shortly after the meeting, which was sent out with a few copies of the book. This report makes it unmistakably clear that I am aware of the desperately shabby, dishonorable treatment I am receiving at the hands of the mathematics community, simply because I dared to be the first to solve, by a short elegant method, a problem that has baffled *your best people* for over 100 years.

Hence, the meaning of my ad should be equally clear. *My proof cannot be refuted.* This ad, *if I must publish it*, makes it necessary for you to show both your students *and me* that *my proof cannot be refuted* (or is it necessary for your students to first show you!?) The international media will help me get this admission from you by publishing a "David challenges Goliath" human interest story about me. I will not be obliged to pay a single cent, let alone $10,000 to anyone, (aside from the relatively small cost of my ad).

I have for some time been in touch with [a newspaper columnist]. I will get all the supplementary publicity I need at no cost,—probably world-wide, through a syndicated feature column.

The latter scenario is therefore divided into two parts:

1. The favorable world-wide publicity I will get when I run the ad—(the world at large looks favorably on the daring underdog),
2. The again favorable world-wide publicity I will receive when you have announced (belatedly) to the media that I have achieved *the world's first short elegant proof of the 4CC*. And contrariwise, the completely unfavorable publicity the mathematics establishment will get when it is realized that you (individually and collectively) have been *essentially guilty of suppression of the truth*.

I don't like the above scenario any more than you do.

I would rather do things amicably, and try to prevent your (individual and collective) loss of face inherent in Part 2 above, by suggesting that you save me the necessity of publishing the ad *by simply announcing to the media as soon as possible* that you, Prof. X. Pert, *have finally come to the conclusion that I have indeed achieved a short elegant proof of the 4CC.* You may even use my new supplement as a face-saving device by claiming that it contains essential ingredients that are necessary to the proof. (The new supplement merely stresses and clarifies what *any moderately intelligent examiner* can find in the original text).

One way or another, I intend to be given *all* the *recognition accolades, academic acclaim, monetary awards, future grants, etc.* that I am entitled to for having *so simply solved* a problem *of such reputed complexity and magnitude.* I trust you will therefore highly endorse the purchase of my book in compensation for the substantial business losses I have sustained due to the form of suppression I have had to endure.

I don't know whether you attended the meeting, and particularly my presentation, but if not, it would still be hard to doubt that you have both examined and discussed my main text, one or more copies of which must have been brought back to the Whatever University by colleagues in your department. Surely you would have made it a point to receive *all details* by some means, since I am sure the AMS Notices and Abstracts would have been sufficient to rouse your curiosity to examine anything presented at this level that could have *any bearing whatsoever* on your work.

I have calculated that, at 40 hours per person per week, taking a count of those who attended my presentation plus an estimation of those with a vested interest to who and with whom my work will have been reported and discussed—*a total man-hour potential of several hundred computer-assisted years has passed since this presentation*, yet:

I HAVE NOT RECEIVED A SINGLE REFUTATION, *because it is impossible to do so.*

Certainly it will be deeply humiliating to you to learn that an amateur has come up with a short elegant solution to this problem of such reputed difficulty, knowing that such a solution has eluded the best of mathematics-oriented minds (including your own) for over 100 years.

The only question is whether you prefer to accept your humiliation as gracefully as possible by giving me the acknowledgement I deserve without further delay, or to let your humiliation be compounded, as will otherwise be the case when you are shown to be among those who are willing to allow this shameful suppression of my work to be continued.

The choice is yours.

It would perhaps be reasonable for you to stall the media by claiming that you need more time to examine the new material, the insert supplement I am enclosing, but if I do not hear, *within a matter of days*, from you and/or other top men in this field to whom I have sent the same request, I will assume that it is *intended* that I should *run the ad, and will do so.* (I would prefer to run the ad *as is*, since it indicates *my complete confidence* that *I will not receive a valid refutation from anyone*. However, if *legal* requirements are such that I must offer the amount to *one person only—the first to comply—and have the offer underwritten* or *otherwise insured*—I will quickly modify the ad accordingly. The one item I will add, in any case, is a *time limit on my acceptance* of *assumed refutations*, since I want to get this business out of the way as quickly as possible.)

I will expect reports, published or confidential, from the student newspapers involved, regarding your reactions, so I trust no attempt will be made to suppress the running of this ad. If such is the case, the media at large will be immediately informed.

You will find in my letter to the Blabbit no mention of suppression, because I do not intend to prove it conclusively unless you refuse my offer, which permits you to announce to the media that the addition of my new material, the insert supplement which you have just examined, finally yields a correct and complete proof. *No suppression can be seen here.*

Sincerely yours,

[W. S.]

In seven years, S. went from being a solid citizen (he owned his own business) with a good mind (judging from his prose) to being a complete crank, with delusions of grandeur and persecution. People are more restrained in their writing than in their thoughts, so imagine the inferno in S.'s head when he wrote his letter! Was his progression inevitable, or could he have been at some point turned aside? His life, I think, has been irreparably damaged and his chance of a serene old age gone. Could he have been saved? The question cannot be answered, but there is a chance that the tragedy could have been averted, though exactly how I do not know.

In 1983, S. was still trying to get his booklet reviewed in mathematical journals. He did not succeed, and I do not know what has happened to him since.

DEDUCTION, THE JOY OF

The genius of Euclid is insufficiently celebrated. When he sat down to write his *Elements of Geometry*, he probably did not know that he was going to create a work that would last for centuries and would forever change how the human race thought, not just about geometry but about truth and how to get at it. He probably did not realize that he was one of the chief figures in a revolution that, in a mere three hundred years or so, made our world possible by elevating reason to its proper place. Probably he was thinking only that it would be a good idea to gather up all the pieces of mathematics that were scattered around (some of them, he may have thought, shockingly badly written) and put them in one place, neatly and logically organized. After all, Euclid worked in a library and may have had some of the habits of mind that librarians are either born with or quickly absorb from the atmosphere of libraries. But whatever his motives, Euclid deserves to be remembered.

Euclid's book has been a model of deductive reasoning ever since it appeared. Deductive reasoning is highly prized as a method of getting at truth, since from truth only truth can follow by deduction. Science is based on it, and even lawyers and politicians attempt to use it at times. It is such a part of our civilization that we cannot imagine a real civilization without it. Deductive reasoning has made us what we are. That is why we insist that high-school students be exposed to a version of Euclid: it is not that any of them will ever have to prove a theorem about triangles, it is just our way of acknowledging our debt to deductive reasoning.

Deductive reasoning can seize the imagination. Spinoza tried to make his philosophical conclusions follow, by deduction, from axioms. Newton presented calculus to the world in the style of Euclid. In 1972, A. R. introduced a new mathematical deductive system, *plerology*. It started, as deductive systems must, with undefined terms:

Primitive Concepts
  Arthroid
  Order
  Junction
  Exterior
  Bounding
  Occupy

1. To each arthroid is assigned a positive integer which is called the *order* of the arthroid.
2. An arthroid may be said to be the *junction of* some set of arthroids.
3. An arthroid or set of arthroids may or may not be *exterior*.
4. If an arthroid or set of arthroids is not exterior it may be *fully bounded*: if it is exterior it is either *partially unbounded* or *fully unbounded*.

R. clearly had in mind some system from which he was trying to abstract the essentials and axiomatize them, to see if deductive reasoning could then give him something that he had not already thought of. Undefined terms can suggest meaning, as "point" and "line" do in Euclidean geometry, and we can get some idea of what will be going on in plerology when we see words like "exterior," "occupy," and "bounded." However, the concept of "arthroid" is not yet clear. Only people completely devoid of mathematical spirit could stop reading:

Defined Concepts

5. An arthroid of order 1 is called a *plere*. An arthroid of order two is called a *bine*. An arthroid of order three is called a *trine*. An arthroid of order 4 is called a *quane*.
6. If an arthroid of order A is the junction of a set S of arthroids, then S is said to *join to form* A. (Symbolically, $S \succ A$ or $A \prec S$.) Any subset S' of S is said to *participate in joining* to form A, and A is said to be the co-junction of S'. Symbolically, $S' \frown A$ or $A \neg S'$.
7. A set of arthroids is said to *fully bound* an arthroid A if and only if $A \frown$ every arthroid in S and the arthroids in S are of order one more than the order of A. If A is not exterior and S contains some but not all of the arthroids, of order one more than the order of A, which A participates in joining to form, then A is said to be *partially bounded* by S. If A is not exterior and if S contains all of the arthroids, of order one more than the order of A, which $A \frown$, then A is said to be *fully bounded* by S. If A does not $\frown$ any arthroid it is said to be *unbounded*.

In a deductive system, we first define our terms, using undefined terms if necessary, and then give the axioms that they satisfy. Plerology has nine axioms, of which the first four are:

Axioms

In the following axioms S and S' will always denote sets of arthroids, while A, A_1, A_2 etc. will denote single arthroids.

I. If $S \succ A$, then the arthroids in S must all be of the same order, and this must be less than the order of A.

II. Given any two positive integers m and n, with $m < n$, every arthroid of order n (n-arthroid) is the junction of precisely one collection of m-arthroids. This collection contains $\binom{n}{m}$ arthroids. (See explanation.)

III. If $A_2 \frown A_1$ and $S \succ A_1$ where the order of A_2 is greater than the order of the arthroids in S, then there is a subset S' of S such that $S' \succ A_2$.

IV. Suppose $S \succ A$ where S is a collection of m-arthroids and A is an n-arthroid. Suppose p is an integer such that $m < p < n$. Then every subset of $\binom{p}{m}$ m-arthroids in S joins to form a p-arthroid which participates in joining to form A.

Now we are ready to go: after the undefined terms, defined terms, and axioms come the theorems, deduced using the laws of logic.

THEOREM A

If a plere participates in forming a trine, it participates in forming at least two bines.

Proof:

Let P_1 be the plere and let T be the trine. By II, T is the junction of precisely 3 pleres. By hypothesis one of these is P_1, and we denote the others P_2, P_3. By IV, $(P_1, P_2) \succ$ a bine which $\frown T$; call this B_1. By IV, $(P_1, P_3) \succ$ a bine which $\frown T$; call this B_2. Therefore $P_1 \frown B_1$ and $P_1 \frown B_2$. (Note: $B_1 \neq B_2$ since by II a bine is the junction of precisely one pair of pleres.)

The next four theorems are:

B: If two pleres do not form a mutual bine they do not form a mutual trine.

C: A plere that participates in joining to form a bounded bine also participates in joining to form some other bine.

D: If two pleres form more than one mutual bine, then no two of those
bines are adjacent.

E: If three bines join to form a trine, each of those bines is the junction
of two of the same three pleres that join to form the trine.

You can see that we have a rich system here, in which any number of theorems
can be proved. To the question of whether the theorems *should* be proved, there are
two answers. The first is: "Sure. Plerology is neat and fun, and I want to see what I
can find out about it." The second is: "No. Who cares? Deduction for its own sake
is sterile and there's no point to it." The first answer reflects the attractiveness of
deduction, while the second comes from the feeling, common among the population
in general, that mathematics ought to be *about* something.

R. explained what plerology is about:

The first part of the word "plerology" is from the Greek root $\pi\lambda\eta\rho$,
which can be transliterated as "pler" with a long "e", pronounced to rhyme
with "here". It means "to fill". The last part is from the familiar $\lambda o\gamma o\varsigma$,
which means "word" or "the thought expressed by a word". Thus plerol-
ogy can be defined as thinking about the filling of space.

In plerology the primary entity is a qualitative space-filler which
we'll call a "plere" and symbolize by P. In some respects it is equivalent
to a "solid" in geometry, though it is arrived at from a different viewpoint.

Think of a particular plere as a lump of red modeling clay. Call it
P_1. Now push P_1 against a lump of blue modeling clay, P_2. The type of
interface where the red clay presses against the blue clay we'll call a two-
junction, or bine. The bine is in some ways the plerological equivalent of
the "surface" in geometry. The particular bine formed by the junction of
the red and blue clay we'll call B_1.

Now picture pressing P_1 and P_2, still pressed together, up against a
pane of glass, and picture what you would see through the glass. The glass
is a third plere, P_3. The locus of the mutual junction of the three pleres
is a three-junction, or trine. In some respects the trine is a plerographic
equivalent of the "curve" in geometry....

Now take a lump of yellow modeling clay, which we'll call P_4. Press
it against the glass and against the red and blue clay in such a way that there
is a common junction among all four pleres. Seen through the glass, it
will look somewhat like the common endpoint of three intersecting curves.
This type of junction we'll call a four-junction, or quane. The quane is in
some respects a plerological equivalent of the "point" in geometry.

That may make the content of the axioms and the theorems more clear.

Deduction is fun, but plerology had a more serious purpose. R. was carrying on the work of F. L., who

> conceived of what is here called plerology as being one of a family of mathematical systems which were to be based on an empirical approach. . . . As one instance he envisaged supplementing the traditional geometries with a metric system related to plerology, in which there would be no such entities as an infinitely small point, an infinitely narrow curve or an infinitely thin surface. Instead, the smallest equivalent of the point, for example, was to be merely very small, not infinitely small. Its size was to be on the order of the threshold of measurability. . . .
>
> He also conceived of eliminating infinite largeness as a concept in his mathematics, since it too has no counterpart in empirical experience. . . . [L.] didn't feel that a non-infinite mathematics should be the only mathematics—just that it would be a useful addition to the tool cupboard of human conceptual techniques, in that it should lead to new and useful applications by virtue of paralleling empirical experience in ways not shared by other mathematical systems.

Thus plerology is not crank mathematics in the sense of being nonsense or demonstrably wrong. What it has in common with crank mathematics is that R. could not get anyone to look at it:

> You're right that I should try to get plerology evaluated by a leading topologist. I tried one, [D.] at [a nearby large university], and a leading algebraic geometer, [A.] at [another large university], but couldn't get either of them to take time.

Being a person of intelligence and good sense, R. did not allow plerology to consume his life, and so he, and it, have faded from the mathematical scene. It may be a loss: an infinity-less mathematics is appealing, and brand-new deductive systems to play with do not come along every day.

Another example of deduction in action has more of the characteristics of crankiness. *The Outline of the Deductive Development of the Theory of the Universe of Motion*, by the International Society of Unified Science, F. M., President, is a 7-page flyer published in 1987 that starts

> Apparently you don't believe us!

After it has your attention, it continues:

> We are a group of scientists, engineers, and others interested in science who are trying to call the attention of the scientific community to a

significant scientific development: D. B. Larson's theory of the universe of motion. What we have been telling you—by "you" we mean the members of the scientific community—is that this theory derives a complete theoretical universe, identical item by item with the observed physical universe, by pure deduction from a single set of premises, without introducing anything from any other source. Obviously, such a development is a major advance in physical knowledge, but it is clear by this time that most of you are not inclined to spend much time and effort examining anything that calls for significant changes in the accepted views of physical fundamentals. So we want to make it easy for you to see that we have something important.

In the accompanying outline we are giving you a small, but substantial, sample of the deductive development. We are stating the premises of the theory, and from them, without introducing anything else, we are making successive deductions, expressed in 54 statements, that account for

A. The *existence* of the following:
  1. Radiation
  2. Matter
  3. Gravitation
  4. The recession of the distant galaxies
  5. The outward motion of massless particles
  6. The "flow of time"
  7. The crystalline state of matter

B. The reason why
  1. Radiation exists in units (photons)
  2. It has both wave-like and particle-like properties
  3. Its speed is independent of the reference system
  4. Matter exists in units (atoms and subatomic particles)
  5. Physical properties are modified at very short distances

This is physical rather than mathematical crankery, but it is very mathematical in deducing the universe from postulates, using only logic. The members of the International Society of Unified Science would be classified, I guess, as neo-Kantians, but they go Kant several better. Kant asserted only that Euclidean geometry gave a priori truths about space; the ISUS asserts that the *entire universe* is a priori! The Society is, by the way, apparently more than a one-person operation: its masthead lists officers (two of them bear the title "Dr.") and an editor of *Reciprocity*, a periodical I have not seen.

Here are the basic premises from which all of the above follow. Euclid had to have five postulates in addition to some common notions, but the ISUS gets along with four:

Basic Premises

The basic premises of the theory consist of certain preliminary assumptions, a postulate, and a definition.

A. In order to make science possible, some preliminary assumptions of a philosophical nature must be made. We assume that the universe is rational, that the same physical laws apply throughout the universe, that the results of experiments are reproducible, etc. These assumptions are accepted by scientists as a condition of becoming scientists, and are not usually mentioned in purely scientific discourse.
B. We assume that the generally accepted principles of mathematics, to the extent that they will be used in this development, are valid.
C. We postulate that the universe is composed entirely of one component, motion, existing in three dimensions and in discrete units.
D. We define motion as the relation between two uniformly progressing reciprocal quantities, space and time.

Motion is all there is. What follows is deduction:

Each of the following statements is a deduction from the postulate and the preceding statements. The objective of the deductive development is to determine what can exist in the theoretical universe defined by the premises of the theory. In most cases it will be evident that the entity or phenomenon that theoretically *can* exist is identical to one that *does* exist in the actual physical universe, and there are no definite conflicts in any case. To the extent that the outline has been carried out, the theoretical universe is thus a correct representation of the observed physical universe.

Since anything that is deduced from something else is implicit in that from which it was deduced, items A, B, C, and D of the premises thus *contain* all of the universe. Inside of them is matter, energy, special relativity, and the expanding universe. There hardly seems room. You would think they would pop from the strain.

The fifty-four things deduced—the theorems of the system—sound like

15. As stated in our definition, motion is a progression. Thus it is not a succession of jumps, even though it exists only in discrete units. There is progression within the units, as well as unit by unit, simply because the unit is a unit of motion (progression). The significance of the discrete unit postulate is that discontinuity can occur only

between units, not within a unit. But the various stages of the progression within a unit can be *identified*.

and

43. When motion takes place in time, the constant progression analogous to clock time is in space, and would be measured by some kind of "space clock." But the rates of progression are the same, one unit of space and one unit of time per unit of motion. Thus the measurements relative to the "space clock" are identical with those relative to a clock that registers time, if expressed in the same units.

The theorems are not in the style of Euclid, and no proofs are given or sketched.

It is an interesting idea that the entire universe is motion, a step up, or perhaps sideways, from Thales's "All is water" and Pythagoras's "All is number." It is a tribute to the ancient Greeks, and to Euclid in particular, that the deductive method has made such a powerful impression that it is still being used to try to get at new truths.

DUPLICATION OF THE CUBE

The duplication of the cube is the least noticed of the three famous problems of Greek geometry unsolvable with straightedge and compass alone—squaring the circle and trisecting the angle are the other two. There is a story that the problem arose because the gods would not call off a plague unless a cubical altar was doubled in size as in Figure 6, but we do not have to pay any attention to that. The problem would have occurred to the Greek geometers without the help of a plague, real or imaginary. Once you have noticed that the diagonal of a square is $\sqrt{2}$ times the length of its side, you have duplicated the square (Figure 7). You can also double the area of any planar figure by magnifying it by a factor of $\sqrt{2}$. Since it is easy to construct a line segment of length $\sqrt{n}$ for any integer n (Figure 8), it is also easy to multiply the area of anything by a factor of n. It is natural then to ask how to increase the volumes of solids by integers. It is also natural to start with a simple solid, a cube, and a small integer, 2. After that problem is taken care of we can worry about more complicated problems, such as how to make dodecahedrons with

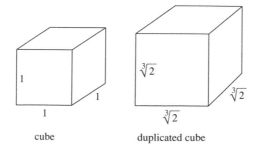

FIGURE 6   cube   duplicated cube

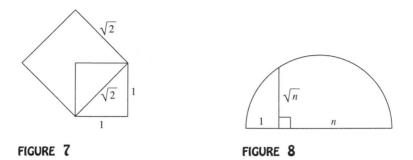

FIGURE 7　　　　　　　　　　　　　　FIGURE 8

volumes five times as large as the ones we have. To duplicate a cube, what is wanted
is a line segment with length $\sqrt[3]{2}$, since magnifying a cube by that factor increases
its volume by 2.

　　Such a line cannot be constructed with straightedge and compass alone. The
ancient Greek geometers no doubt realized this, so they found solutions using other
means. If we can find numbers x and y so

$$\frac{1}{x} = \frac{x}{y} = \frac{y}{2},$$

then

$$2 = \frac{y^2}{x} = \frac{x^4}{x} = x^3,$$

and we will be able to duplicate the cube. That is, if we can locate the intersection
of the parabolas

$$x^2 = y \qquad \text{and} \qquad y^2 = 2x,$$

then we are done (see Figure 9).

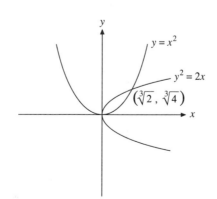

FIGURE 9

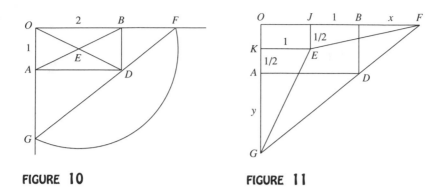

FIGURE 10   FIGURE 11

Around 220 B.C., Apollonius gave another solution: take a rectangle $OBDA$ with sides of length 1 and 2, and draw a circle with center at E that cuts the extended sides of the rectangle so that F, G, and D lie on a straight line (see Figure 10). This construction is not Euclidean: to do it, you need to take your straightedge and wiggle it around D as a pivot until F and G are equally distant from E. The wiggling straightedge is not a Euclidean tool. When the construction is done, BF has length $\sqrt[3]{2}$ and thus is the edge of a cube with volume 2.

It is not hard to see why the construction works. In Figure 11, triangles BDF and AGD are similar, so

$$\frac{x}{1} = \frac{2}{y} \qquad \text{or} \qquad xy = 2.$$

Applying the Pythagorean theorem to triangles KEG and JEF gives

$$r^2 = 1^2 + \left(y + \frac{1}{2}\right)^2 = \left(\frac{1}{2}\right)^2 + (1 + x)^2,$$

so

$$1 + y^2 + y + \frac{1}{4} = \frac{1}{4} + 1 + 2x + x^2$$

or

$$y^2 + y = 2x + x^2.$$

Multiplying by x^2 and using $xy = 2$ yields

$$4 + 2x = 2x^3 + x^4,$$

so

$$0 = x^4 + 2x^3 - 2x - 4 = (x^3 - 2)(x + 2).$$

Since $x \neq -2$, it follows that $x^3 = 2$.

Other solutions were given in antiquity, and in more modern times: Huygens, Viète, Descartes, and Newton all addressed the problem.

Cranks, on the other hand, have not paid much attention to it. Some throw in a solution to round out their trisections of the angle and squarings of the circle, but you can tell that their hearts were not in it as much as they were for the two bigger problems. For example, there is D. J., author of *Solutions of the Three Historical Problems by Compass and Straightedge*, a hardbound book that a vanity publisher took J.'s money to publish. The first part of the book is devoted to dividing angles into 3, 5, 7, and other parts, then comes the quadrature of the circle, and it is only toward the end of the book that we have

Theorem:

A given cube can be doubled in volume if the area of the sides of the cube is doubled.

What that seems to say—what that *does* say—is that the volume of a cube whose faces have area 2 is twice that of a cube whose faces have area 1. However, the edge of a cube whose faces have area 2 is $\sqrt{2}$, so its volume is $(\sqrt{2})^3 = 2.828427125\ldots$, a long way from 2. J. was not really trying hard when he produced that result. Here is the proof of his theorem, in its entirety:

An inscribed square in a circle and an inscribed cube in a sphere of the same diameter portray the square and a side of the cube are equal. Both are holding the same position in equal circles. It is obvious that doubling a square and doubling a cube are one and the same operation. Doubling the feature of a square and doubling the feature of a cube being two problems, is geometrically a misconception.

I think that we have here a case of failure to understand the problem. It is not a case of rushing into print, since I have a copy of the privately published pamphlet that preceded J.'s book, and it has copyright dates of "1962–1963–1965–1968–1971–1972." When I wrote J., making some polite inquiries about his work, I got no reply, which is unusual for cranks. I hoped that it was because he had given up on the duplication, trisection, and quadrature, but it was not so, as the vanity press volume shows. It may have been that he had had enough of communicating with mathematicians.

Someone should write a history of the office of Notary Public, to satisfy my curiosity. Notaries are clearly holdovers from a former day, and I would like to know when and how the need for them arose (medieval Europe was when, most likely, but how the idea of notary first occurred to someone is a mystery to me), when they reached their peak of power and influence, and what caused their decline. Surely there were great notaries: who were they and what were their heroic deeds? Were there corrupt and evil notaries, abusing their trust? Questions like these too often go unanswered.

What brings notaries public to mind is a 1971 letter by W. V., addressed

TO WHOM IT MAY CONCERN

It starts:

> Sirs, I am hopeful that someone at Lloyds of London will inform me if there are odds set on someone solving the famous problem of antiquity known as the "Duplication of The Cube".

It later states:

> The Encyclopedia Britannica states that this problem cannot be solved by the use of a straight edge and compass alone, and many famous authoritative geometricians and mathematicians firmly agree with this statement. I challenge them all.
>
> Since I am challenging all past, as well as present authorities that are in agreement with the Encyclopedia Britannica the turmoil this should cause could bring some interesting results. If the odds appeal to me, I will be most happy to reveal this knowledge to all that challenge me.

I have three copies of this letter, but all at second hand, so I do not know if anyone took up the challenge nor what the details of the duplication were. Another approximate duplication lost to history! Be that as it may, the letter at the bottom was "Subscribed and sworn to, before me, this 10th day of February, 1971 at Detroit, Michigan" and signed by a Notary Public of Oakland County, Michigan, whose commission expired on August 10, 1973. (*What* commission? And why are notaries compelled to include their dates of expiration on everything that they notarize?) It seems to be a tenet of popular culture that important documents should be presented to a Notary Public as a sign of their importance, just as popular culture holds that Lloyd's of London is in the business of setting odds on unlikely events and that the *Encyclopedia Britannica* has the responsibility of ascertaining and setting forth truth. What was going through V.'s mind as he presented his letter to the notary? What was going through the notary's mind as he scrunched his seal on it? These are questions as impossible to answer as the cube is impossible to duplicate.

Now for a real duplication. *Duplicating the Cube* by C. T. is a nicely printed nine-page pamphlet, copyrighted 1981. T. said:

Obtaining an exact measurement of the cube root of two (an "irrational" length) with straight edge and compass, and prove the device "constructible" is the name of the game. . . . The following argument applies a new (or forgotten) method to the old problem in order to avoid the usual difficulties.

Then he plunged into his construction. It is not quite Euclidean since it uses a marked straightedge, the same tool that Archimedes used to trisect the angle. The construction is easy to do and easy enough to verify to provide valuable exercise to students of geometry and trigonometry. Here it is (Figure 12): draw a 30-60-90 degree triangle OAB and a semicircle with the hypotenuse for its diameter. Then take a straightedge with the distance from O to B marked on it and draw a line through B so the distance from D, on the semicircle, to E, on a line perpendicular to OA, is the marked distance. Then $|AE|$ is the line you need to duplicate the cube: it has length $\sqrt[3]{2}$. Is that not a pleasing construction, and, if it is original with T., was he not amazingly ingenious to find it? The answers are yes, and yes.

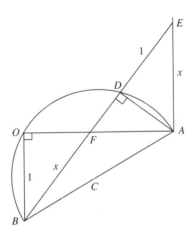

FIGURE 12

To see why AE has the right length, notice that angle BDA is a right angle, as are all angles inscribed in semicircles by starting at the ends of a diameter. Thus triangles ADE and FAE are both right triangles. Since they have the same angles (one angle 90° and angle AED in common, so their third angles are the same) they are similar. So, if we let $|DE| = 1$, we have

$$\frac{|DE|}{|AE|} = \frac{|AE|}{|EF|} \qquad \text{or} \qquad \frac{1}{x} = \frac{x}{|EF|},$$

so $|EF| = x^2$. There is another pair of similar right triangles, FEA and BOF, so

$$\frac{|BO|}{|BF|} = \frac{|AE|}{|EF|} \qquad \text{or} \qquad \frac{1}{|BF|} = \frac{x}{x^2},$$

and $|BF| = x$. Since $|OB| = 1$, the hypotenuse of triangle BOA has length 2, so $|OA| = \sqrt{3}$. Thus

$$\sqrt{3} = |OF| + |FA|$$
$$= \sqrt{x^2 - 1} + \sqrt{x^4 - x^2} = \sqrt{x^2 - 1} + x\sqrt{x^2 - 1}$$
$$= (1 + x)\sqrt{x^2 - 1}$$

or

$$3 = (1 + 2x + x^2)(x^2 - 1) = x^4 + 2x^3 - 2x - 1,$$

so

$$x^4 + 2x^3 - 2x - 4 = (x + 2)(x^3 - 2) = 0.$$

Since $x \neq -2$, we have $x = \sqrt[3]{2}$.

The equation is the same as the one that arose in Apollonius's duplication, and both constructions contain right triangles with side of length 1 and hypotenuse of length 2, so they may be equivalent. If so, it is probably coincidence. The Archimedean trisection has been rediscovered more than once, so it is not impossible for the same thing to happen to the Apollonian duplication.

Next you could ask if the method of Apollonius or T. could be used to *triple* the cube. The answer is yes.

By the way, rewriting $\sqrt{3} = |OF| + |FA|$ with the value of x substituted, we get the true statement

$$\sqrt{\sqrt[3]{4} - 1} + \sqrt{\sqrt[3]{16} - \sqrt[3]{4}} = \sqrt{3},$$

which, presented in that fashion, is not at all obvious. Anyone who can quickly verify that it is correct using algebra alone is someone with considerably more mathematical talent than the average person.

ELLIPSE, CIRCUMFERENCE OF AN

The problem of *circumambulating the ellipse* has never had the same appeal as that of squaring the circle, even though they are similar. To square the circle is to determine the value of π, the ratio of the circumference of a circle to its diameter, while to circumambulate the ellipse is to determine the circumference of an ellipse in terms of the lengths of its major and minor axes. The reason for the circle's dominance is that everyone knows about circles: the ancient Greeks worked on the problem, everyone runs into π at an early age, and teachers often talk about circles. Ellipses, on the other hand, hardly penetrate the consciousness of students today, and even when they do there is no strange symbol like π to pique the interest. Teachers do not mention the problem of finding the circumference of an ellipse because it leads to integrals that students cannot evaluate.

Given an ellipse, say

$$x^2 + \frac{y^2}{4} = 1,$$

all students of calculus should know that to find its arc length s we first solve for y,

$$y = 2\sqrt{1 - x^2}$$

(we need only look at one quadrant of the ellipse), to get

$$dy = 2 \cdot \frac{1}{2}(1 - x^2)^{-1/2}(-2x)dx = \frac{-2x}{\sqrt{1 - x^2}}dx.$$

Then, of course, we approximate infinitesimal arc lengths using the Pythagorean theorem. Substituting our result from above, we get

$$(ds)^2 = (dx)^2 + (dy)^2 = (dx)^2 + \frac{4x^2}{1 - x^2}(dx)^2,$$

from which we get

$$(ds)^2 = \frac{1 + 3x^2}{1 - x^2}(dx)^2.$$

Thus

$$s = 4 \int_0^1 \sqrt{\frac{1 + 3x^2}{1 - x^2}}\, dx$$

gives the arc length of the ellipse—the factor of 4 takes us all the way around, top and bottom.

The integral looks no more difficult than many that appear in calculus textbooks, but no student of calculus could evaluate it without making a mistake. The natural substitution, $x = \sin\theta$, leads to

$$s = 4 \int_0^{\pi/2} \sqrt{1 + 3\sin^2\theta}\, d\theta,$$

which looks even more like something that would be right at home in a list of problems in a textbook, but nothing can be done with it. It is an elliptic integral of the second kind, not to be dealt with by students beginning the study of calculus and not to be integrated in closed form in terms of elementary functions by anyone. Thus the circumference of an ellipse does not get mentioned in classrooms, and therefore cranks do not make it their life's work to determine it.

What a shame, since the circumference can lead to all sorts of activities of more or less fascination. For example, expand the last integrand as a series, integrate term by term, and see what happens. Duplicate the arc length calculation for the general ellipse

$$\frac{x^2}{a^2} + \frac{y^2}{b^2} = 1,$$

and show that as a approaches b we get the arc length of a circle. Show that as b/a approaches infinity we get the arc length of a parabola. Define the *diameter* of an ellipse to be that number such that the ratio of the ellipse's circumference to the diameter is π, and investigate the relation of the diameter to a and b. Or, define

$\pi(a, b)$ to be the ratio of the circumference of an ellipse with semiaxes a and b to $a^2 + b^2$, or to $\sqrt{a^2 + b^2}$, or to its eccentricity, and investigate the diameter's properties. Mathematics, even at its elementary levels, is endlessly rich.

The field of crankery is almost as rich. Given almost any problem that cannot be solved, a crank will arise to solve it. An undated sheet by J. W. is headed

FIRST MATHEMATICAL FORMULA FOR CIRCUMFERENCE OF ELLIPSE

and starts:

> The following formula represents an important breakthrough in the mathematics necessary in astrophysics and other advanced sciences. The ellipse is the characteristic pattern of the orbit of a satellite and other astral bodies. The lack of a mathematical formula for calculating the circumference of the ellipse has retarded advances in these fields.

In W.'s picture (Figure 13), the foci are at the ends of the segment with length C, and E is the arc length of the bottom half of the ellipse, half of the length we would like to know. Here is how we find it:

> Find the eccentricity of the ellipse. The eccentricity is the distance between the two foci divided by the length of the major axis. Eccentricity equals C/M.
>
> Use the enclosed Graph Relating The Eccentricity To Slope to find the slope. Values for the slope are listed on the left side of the graph. Values for eccentricity are on the bottom.
>
> The slope is the distance between the foci divided by one half the circumference of the ellipse. Slope equals C/E.
>
> Divide the distance between the foci by the slope to find one half of the circumference. Double to find the circumference.
>
> The first mathematical formula for finding the circumference of an ellipse is:

$$\text{Circumference} = \frac{2C}{S}$$

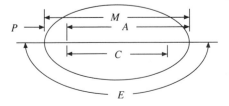

FIGURE 13

Because the slope was defined to be C/E, the formula says that the circumference is

$$\frac{2C}{C/E} = 2E,$$

which is a tautology. Of course, if we could calculate the value of S from known quantities we would have a real formula, but W. would have us read it off his graph (Figure 14). So, W.'s method amounts to no more than an approximation and thus has no advantages over other approximate methods for finding the circumference.

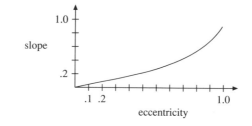

FIGURE 14

W.'s work is nevertheless interesting. The idea of the slope of an ellipse was a new one to me ("slope" is not the right word—what is?) and it is interesting that its graph against eccentricity is nearly linear for eccentricities up to $1/2$. Surely a quadratic or cubic could be found that would approximate the curve *very* closely, and then we could get a handy formula for determining approximate circumferences of ellipses very quickly. It is not that there is any great need for such a formula, handy or otherwise, but finding such things can provide innocent amusement.

ENCOURAGING CRANKS, THE FOLLY OF

Cranks can escalate, as a 48-page file of material from G. D. illustrates. D.'s initial letter claimed to prove that any factor of a Mersenne number,

$$2^p - 1, \qquad p \text{ prime}$$

must have the form $66k \pm 1$. No teacher of mathematics could resist gently pointing out one of the many counterexamples, such as

$$2^{11} - 1 = 2047 = 23 \cdot 89,$$

where 23 is obviously not of the form $66k \pm 1$. D. then scaled his claim down to $2kp \pm 1$, which in fact is true, but he also asserted that $2^{29} - 1$ is prime. No teacher of mathematics could let an error like that go uncorrected, so a reply was necessary pointing out that

$$2^{29} - 1 = 536870911 = 223 \cdot 1103 \cdot 2089,$$

together with a condescending remark or two about how easy it is to miss large factors like 223, especially when calculators can hold only eight digits.

But then the letters started to flow, containing philosophy:

> By the very nature of knowledge, and of man, man is and may become nothing more or less than what he accepts makes of him.

Even worse, the letters included more and more new results:

I shall send you my proof of the extended F. L. T. viz., "If

$$a^n \pm b^n \pm c^n \pm d^n = 0, \tag{1}$$

is to be satisfied in integers, then n must be one of the first five numbers, i.e. $n = 1, 2, 3, 4,$ or 5."

The classic use of the Theory of Least Squares weights a scientific skewing into the data which makes true results quite impossible.

The five postulates of Euclid for his plane geometry are mere corollaries of the axiom, used as postulate number III by Euclid—two points in a given plane always uniquely define one and only one circle just as soon as assignments of center point and circle point are made.

The four color problem is solved by accepting the proposition,— "five points can mutually connect each other by lines only if at least two lines intersect each other".

Poincaré's Theorem merely states,—"One must use proper tools if one is to be rational in a given analysis".

On and on the new theorems came. D. was generous: one of his letters concluded:

If you have a paradox you desire solved, please let me have it.

On and on the new letters came. Bitterness welled up:

I fully realize the sycophant reverence the classic scientist holds for the erudite dogmas of mystery called,—"absolute-rigor," and therefore do not ask you to comment on the following formula giving an optimum solution of the elliptic integral, ...

Eventually, after a sufficient number of non-replies, the correspondence came to an end.

D. D. was a crank who responded to attention. Not knowing his character, I once wrote him a note expressing surprise at a numerical coincidence that he had found. The result was that in the next month or so I received at least ten multi-page letters from him, to none of which (for I was beginning to understand his character) did I respond.

The coincidence, by the way, is this. Take the six permutations of 1, 2, and 3, arrange them in increasing numerical order, and take the differences between them:

123		132		213		231		312		321
	9		81		18		81		9	

The sums of the two lines are 1332 and 198. (The first number provides a reason for the number of words in the Declaration of Independence missed by A. F., the investigator of the American Revolution.) Now take the first nine digits of π and add 198 three times:

$$
\begin{array}{ccc}
314 & 159 & 265 \\
\underline{198} & \underline{198} & \underline{198} \\
512 & 357 & 463
\end{array}
$$

Add: $512 + 357 + 463 = 1332$. That is a striking coincidence.

I should not have told D. that his calculation was striking because he then thought so too, and he included it prominently in material that he sent to other people.

I have what seems like a ton of material by and from this vigorous crank, and some of it shows the reason why there is so much. Before getting to that, let me give a sample of D.'s work so it can be seen that it does not warrant encouraging. A sample will do, because almost all of the ton is indistinguishable from it. D. wrote:

> I am sending on a paper which I recently obtained copyright on which centers on the terminating decimal $1/3$. Thank you and *phease* let me know if this letter is received.

Yes, "terminating decimal $1/3$" is what he wrote. He had invented a new symbol looking like $\leftrightarrow$ that meant, he wrote, "inverts to." Here is his page on $1/3$:

(1) $1/3 = 0.333$ or $333 \leftrightarrow 666$

(2) $333 \div 666 = 2!$

(3) $2 \leftrightarrow 7$ or $2 \div 7 = 35 \leftrightarrow 64$ or $64 \div 35 = 05!$

(4) $05 \leftrightarrow 94$
$$
\begin{array}{r}
- 05 \\
\hline
89^2 = \underline{7921}
\end{array}
$$

(5) $2 \div 35 = 175 \times \pi/4 = \underline{700}!$

(6) $7921 \div 700 = 00883726 \leftrightarrow 99116273$

(7) $700 \div 99116273 = 01415946$—NOTICE!

(8) $01415946 \leftrightarrow 98584053 \div$ by $2 = 492920265$

(9) $2 \div 242920265 = 2464601325 \leftrightarrow$
$7535398674 \times 4 = \pi/4!$
$\pi = 30.141594696$

(10) $1/3 = 333 \overline{)30.14159469600000}$ over which is 0.090515299387387387—repeats!
$\therefore 1/3 = \pi = 30.141594696!$

Clearly, "inverts" means take the nine's complement, but equally clearly the equals sign meant something different to D. than it does to the rest of the world. In (5) we have $\pi = 16$, and in (9) it equals something much larger. I assure you that I am not selecting unfairly: this page is typical of D.'s work. The first line of a letter to an eminent mathematician at a prestigious university (how D. found out to whom he should write is another mystery) began:

> The "Biblical" value of $\pi = 30/10 = 3$ can be *proven* that $3 = 3/8$
> which reads:

The letter continued with more numbers filling the page.

D. sent me copies of some letters he had received, and they do much to explain why he continued to bombard the mathematical world with his utterly incomprehensible equations. The managing editor of *The Mathematics Teacher*, official journal of the National Council of Teachers of Mathematics wrote (after misspelling D.'s name):

> I am pleased that you have found some success in your work. You might wish to excerpt some small portion of the material and send it to one of the journals listed on the enclosed sheet.

That got rid of the nuisance of D., but writing that letter is just shameful. I will not say when the letter was written, so the managing editor cannot be identified; but I can say that he acted irresponsibly in giving D. a list of other places to send his material. The managing editor should have been able to recognize crank mathematics—a managing editor of a mathematics journal does not deserve his job if he cannot— and he should have recognized his duty, both to the crank involved and to the mathematical community, not to encourage cranks.

Here is another shameful example. An employee of the International Business Machines Corporation, a holder of the PhD degree, wrote to D.:

> It is entirely possible that some publisher would be willing to publish your material. I wish you luck.

It is possible that the PhD knew no better and could not distinguish nonsense from sense. It is more likely that he was behaving dishonestly, encouraging someone to continue in an effort that he knew was doomed to failure.

Another person who put a "Dr." in front of his name wrote to D.:

> I have enjoyed immensely your mathematical formulas. You have done a tremendous amount of work.
>
> Your mathematical genius and perspectives certainly warrant recognition. I am not sure what steps you should take but my feeling is that you might proceed by thinking in terms of a book.

This advice may not be as irresponsible as that given earlier since the person who gave it had written a book on the theory and applications of pyramid power. People who believe in pyramid power cannot be held to standards as high as those to which managing editors should adhere. On the other hand, the pyramid-power doctor probably did not have much mathematical training, and people who lack mathematical training should neither encourage nor discourage mathematical writers.

This sort of thing goes on and on. Government agencies must by rule be polite to the public, so it is no surprise that a letter to D. from the Lyndon B. Johnson Space Center included:

> I very much appreciate receiving your letter of [recent date] with the enclosed letter of evaluation by [the pyramid doctor above]. You are to be congratulated for your fine work and success in receiving evaluations by Drs. [Pyramid and IBM].

However, it is possible to be polite without encouraging crankery. When you do not understand something, the proper course is to admit it. The proper form letter, typed into the office word processor and retrievable with a keystroke or two, should read something like the following:

> Dear [fill in the space]:
>
> We have received the material you sent with your letter of [fill in this space too].
>
> We regret that no one here is competent to pass judgment on its quality, so it is herewith returned.
>
> Sincerely yours,

This letter is for the case in which there is in fact no one who can recognize crank mathematics. When there is such a person, as there should have been at *The Mathematics Teacher*, the form letter should read something like this:

> Dear [crank]:
>
> Your paper, [fill in title], is not suitable for publication in [this journal].
>
> Work such as you have done can provide enjoyment to the person who does it, but that will be its only reward. I suggest that you do not seek to have it published elsewhere.
>
> Sincerely yours,

Accurate, honest, and appropriate. More letters like this should be written.

EQUATIONS, SOLVING

Everybody learns that if

$$ax^2 + bx + c = 0,$$

then

$$x = \frac{-b \pm \sqrt{b^2 - 4ac}}{2a}.$$

In the old days, when courses called "Theory of Equations" were taught, students learned formulas that solve, exactly,

$$ax^3 + bx^2 + cx + d = 0,$$

and even

$$ax^4 + bx^3 + cx^2 + dx + e = 0.$$

In advanced courses, students learn that there is no formula for solving

$$ax^5 + bx^4 + cx^3 + dx^2 + ex + f = 0,$$

nor are there any for polynomial equations of higher degree, and they learn the reason why.

You would think that when teachers say, "It is impossible to find formulas to solve polynomial equations of degree five or higher," it would be as provocative as "It is impossible to trisect angles with straightedge and compass alone," but the number of people who try to solve fifth-degree polynomial equations is tiny compared with the number who try to trisect angles. Maybe the problem is not as appealing, maybe teachers do not mention it very much, or maybe it is harder for a crank to convince himself that he has solved fifth-degree equations than it is for him to convince himself that he has trisected the angle.

Since the number of businesses that solve mathematical problems is also tiny, it is surprising that it was a corporation in Illinois that sent out this advertisement in 1989:

Re: Marketing of solutions to equations

What: Is marketing solutions to equations such as:

$$x^3 + ax^2 + bx + c = 0$$
$$x^4 + ax^3 + bx^2 + cx + d = 0$$
$$x^5 + ax^4 + bx^3 + cx^2 + dx + e = 0$$
$$x^6 + ax^5 + bx^4 + cx^3 + dx^2 + ex + f = 0$$
$$x^7 + ax^6 + bx^5 + cx^4 + dx^3 + ex^2 + fx + g = 0$$
$$x^8 + ax^7 + bx^6 + cx^5 + dx^4 + ex^3 + fx^2 + gx + h = 0$$

When: Now available exact solution to equations not approximations

Why: The possibilities are many. Some of the possibilities are:
   * To regain world leadership in the technologies.
   * To understand nature's method of control.
   * For solutions to nuclear controls—possibly GUT theory. There-fore an acceptance of nuclear power generation by the public.
   * For possible nuclear waste management.
   * For solutions to chemical bonding as in pharmaceutical re-search. Also to catalogue chemicals by mathematical equations.
   * The possibility of controlling variables $(a, b, c, \ldots)$ in powerful and power generating formulas. Also propulsion fuel genera-tion.
   * The possibility of controlling never before events, and under-standing how these equations control nature, man-made events, and results.
   * To develop energies beyond nuclear power as is being researched at M.I.T.
   * The possibilities are endless.

What a tribute to the power and majesty of mathematics! The subject and discipline have such force and splendor that people get the idea that mathematics has the ability to do wondrous things when in fact its range is limited. Even if seventh-degree equations could be solved exactly in an instant, we would be no further with nuclear waste management. Even when the corporation solves seventeenth-degree equations (it is working on them), chemical cataloging will not be revolutionized. It would be pleasant if solving some equations could solve the problems of the world, but that is not the nature of the world, or of equations.

My letter to the corporation got no response, probably because it contained downbeat sentences like

> You may find that no one is going to pay your claims very much attention. The reason is that it has been known for 170 years that it is impossible to solve those equations in general.

I should have replied, "Yes! Send me details of your exciting solutions!" but the honesty that is endemic among mathematicians prevented me. Thus I will never know how the corporation proposed to do the impossible.

FERMAT'S LAST THEOREM

The most famous equation in mathematics, far ahead of the runner-up

$$x = \frac{-b \pm \sqrt{b^2 - 4ac}}{2a}$$

($E = mc^2$ does not count because it is a physics equation), is

$$x^n + y^n = z^n.$$

This is the equation that Pierre de Fermat (1601–1665) wrote about in his famous marginal note in his copy of the works of Diophantus, next to problem 8 in book II, "Given a number that is a square, write it as a sum of two other squares":

> On the other hand, it is impossible for a cube to be written as a sum of two cubes or a fourth power to be written as a sum of two fourth powers or, generally, for any number that is a power greater than the second to be written as a sum of two like powers. I have a truly marvelous proof of this proposition that this margin is too narrow to contain.

That is,

$$x^n + y^n = z^n$$

has no solution in positive integers if $n > 2$.

This has become known as "Fermat's Last Theorem," a double misnomer if ever there was one. First, it isn't a theorem since no one has proved it yet and it may be false. Second, it wasn't Fermat's last anything: though exactly when he wrote it is unknown, it was probably relatively early in his mathematical career, at some time around 1637. Better it should be called "Fermat's Conjecture" or "One of Fermat's Early Wrong Guesses"; but usage is what defines language, and "Fermat's Last Theorem" is well established. So, FLT it will have to be.

FLT has had a great effect on modern mathematics. Many people believe that mathematics makes progress by mathematicians' efforts to solve problems that arise in the world external to mathematics. Fewer people, but still many, believe that the positive integers exist quite independently of mathematics and, for that matter, of the human race, and so trying to prove FLT is analogous to attacking a problem in physics. Be that as it may, attempts to prove FLT have advanced mathematics. It is because of FLT that all majors in mathematics (well, almost all) memorize the definition of "ideal". Without FLT we would know less about class numbers and cyclotomic polynomials. It can be argued that FLT *caused* algebraic geometry. In any event, it changed the course of mathematical history.

The question arises, was FLT inevitable? A case can be made that it was not. If Fermat did not have a copy of Bachet's edition of Diophantus's works, he would not have had a margin to write it in. If Diophantus had not written his works, Fermat might never have thought about the problem of writing integer powers as sums of two integer powers. And Diophantus was a fluke of history if ever there was one. He lived and worked in Alexandria at some time from 150 to 350 A.D. (no one knows exactly when), and he is unique in the annals of Greek mathematics. The ancient Greeks were geometers and did no algebra. None of them even thought about equations, much less about solving them, until Diophantus. Diophantus, however, did. Diophantus also required that solutions of his equations be rational numbers. Again, the reason is unknown. His writings give no reasons for wanting to solve the problems he considers before he plunges in to show how to find, for example, a square that, when increased or decreased by five, is again a square. Diophantus was, I think, an Egyptian, hellenized, very smart, aware of the Egyptian mathematical tradition, fascinated with mathematics, and lucky to have an academic job in Greek society. Any person who satisfies all six conditions qualifies as a fluke. Diophantus had neither predecessors nor successors, so if he had never existed the problems that he considered would not have been thought of in his time or for centuries after. The problems might then *never* have been thought about, since they do not arise in physical reality. Thus there would be no FLT, no books called *Rings and Ideals* would ever be written, and hundreds—thousands—of cranks would have had to find something else to be cranky about.

FLT would probably have come up, though. Pythagorean triples, solutions in integers of $x^2 + y^2 = z^2$, go back a long way, and it does not take a terribly agile mathematical mind to go from them to asking about solutions of $x^3 + y^3 = z^3$ in integers, and then to solutions of $x^n + y^n = z^n$. The question is, would anyone have paid attention if some minor writer had brought up the equation? There have been enough such cases—papers written and forgotten and only recognized years later as major discoveries—to make it plausible that without the immense reputation of Fermat behind it, progress on FLT would have been slower or nil. We will never know, though I suspect that the obvious appeal of the problem would eventually have attracted the attention of someone else with an immense reputation.

Also, the proof that there is no solution for $n = 4$ is simple enough that it could have been found by someone other than Fermat, and that would have been a great impetus to attack the problem further. The proof is actually rather natural since it parallels the process by which $x^2 + y^2 = z^2$ is solved. To solve that equation, it is natural to write

$$x^2 = z^2 - y^2 = (z - y)(z + y),$$

then say that the product of two numbers is a square if they are both squares,

$$z - y = b^2 \quad \text{and} \quad z + y = a^2.$$

Thus,

$$z = \frac{a^2 + b^2}{2}, \quad y = \frac{a^2 - b^2}{2}, \quad \text{and} \quad x = ab.$$

Put $a = 3$ and $b = 1$, and you have the well-known solution $(x, y, z) = (3, 4, 5)$; put $a = 27$ and $b = 17$ and you have another solution, $(x, y, z) = (459, 220, 509)$:

$$459^2 + 220^2 = 210681 + 48400 = 259081 = 509^2.$$

In fact, all solutions have been found, a great success.

After that success, it would be natural to attempt repeating it with $n = 4$: from $x^4 + y^4 = z^4$ we get

$$x^4 = z^4 - y^4 = (z^2 - y^2)(z^2 + y^2).$$

To have the product of two numbers equal to a fourth power we can try

$$z^2 + y^2 = a^2 \quad \text{and} \quad z^2 - y^2 = b^2,$$

with $ab = c^2$. Thus $2y^2 + b^2 = a^2$, so

$$2y^2 = a^2 - b^2 = (a - b)(a + b).$$

This can happen if one of the factors is a square and the other is twice a square:

$$a + b = 2r^2 \quad \text{and} \quad a - b = s^2.$$

Thus,

$$a = \frac{2r^2 + s^2}{2}, \qquad b = \frac{2r^2 - s^2}{2}, \qquad \text{and} \qquad y = rs.$$

Now we have y, so we can substitute in $z^2 = y^2 + b^2$ to get z:

$$z^2 = r^2 s^2 + \frac{4r^4 - 4r^2 s^2 + s^4}{4}$$

or

$$4r^4 + s^4 = 4z^4.$$

The trouble is, this does not quite give z, as happened when n was 2; rather, it gives an equation for z^4. Furthermore, the equation looks no simpler than the one we started with,

$$x^4 + y^4 = z^4.$$

We have not made much progress.

In fact, we cannot make any progress at all. When the preceding idea is refined and polished, it becomes the method of infinite descent and it can used to prove that the equation has no solutions in positive integers. The method consists of assuming that the equation has a solution in positive integers, then from that solution deducing the existence of another solution in *smaller* positive integers. That is, from one solution we can descend to a smaller one. From the second solution we can descend to a third, a fourth, and so on: an infinite descent through positive integers. That of course is impossible—we will bump into zero sooner or later—so the assumption with which we started, that there is a solution in positive integers, must have been wrong.

A proof of FLT for $n = 3$ was given by Euler in 1770. It too used the method of infinite descent, and it was considerably more complicated than the proof for $n = 4$. In fact, it had a flaw, though one that could be and was repaired. Dirichlet and Legendre settled $n = 5$ in 1825, and Lamé took care of $n = 7$ in 1839. In 1847, Kummer proved the theorem for an entire class of integers, and work and progress have continued ever since.

The theorem might never have attracted the attention of amateurs. It is true that it is easy to state and understand, but so are Goldbach's conjecture that even integers are the sum of two primes and the twin prime conjecture that there are infinitely many pairs of prime numbers p and $p+2$. While some amateurs think they have solved those problems, their number is nothing compared to the multitudes who think they have proved FLT. The Riemann hypothesis, that the zeta function has zeros in the critical strip only for $s = 1/2$, is also easy to state, but *no* amateur bothers with it.

The reason FLT is so popular is the dratted Wolfskehl Prize. Somehow the eye of Dr. Paul Wolfskehl was caught by the FLT (that Fermat's *Works* were published in Paris in 1891 may have had something to do with it), and when he died in 1908 his will provided a prize of 100,000 marks to be given to the first person to prove the theorem. This substantial sum attracted the attention of many people, and ever since it has been folklore that fortune and fame await whoever can prove FLT. The prize is now worth only $10,000 or so, but word of that has not gotten around. Even after the prize goes out of existence—which it is scheduled to do if it is not awarded by September 13, 2007—proofs will no doubt continue unabated. The main result of Dr. Wolfskehl's prize has been to add to the total of human unhappiness, which is not what he intended.

The prize created the unfortunate tribe of fermatists. "Fermatist," a term that is not original with me, though I have forgotten where I first saw it, is a handy abbreviation for "a person who thinks that he or she has proved Fermat's Last Theorem." It will be used hereafter when convenient. No similar terms need to be invented for other classes of cranks because none are needed: "trisector" and "circle-squarer" are short and descriptive, and there are too few duplicators of the cube to justify introducing "cubist."

In *13 Lectures on Fermat's Last Theorem*, Paulo Ribenboim quoted a 1974 letter from Dr. F. Schlichting of Göttingen about fermatists and the Wolfskehl prize, which says, in part,

> There is no count of the total number of "solutions" submitted so far. In the first year (1907–1908) 621 solutions were registered in the files of the Akademie, and today they have stored about three meters of correspondence concerning the Fermat problem. In recent decades it was handled in the following way: the secretary of the Akademie divides the arriving manuscripts into (1) complete nonsense, which is sent back immediately, and (2) material which looks like mathematics. The second part is given to the mathematical department and there, the work of reading, finding mistakes and answering is delegated to one of the scientific assistants (at German universities these are graduated individuals working for Ph.D. or

habilitation and helping the professors with teaching and supervision)—
at the moment I am the victim. There are about 3 or 4 letters to answer
per month, and there is a lot of funny and curious material arriving, e.g.,
like the one sending the first half of his solution and promising the second
if we would pay 1000 DM in advance; or another one, who promised me
10 per cent of his profits from publications, radio and TV interviews after
he got famous, if only I would support him now; if not, he threatened to
send it to a Russian mathematics department to deprive us of the glory of
discovering him. From time to time someone appears in Göttingen and
insists on a personal discussion.

Nearly all "solutions" are written on a very elementary level (using the notions of high school mathematics and perhaps some undigested
papers in number theory) but can nevertheless be very complicated to understand. Socially, the senders are often persons with a technical education
but a failed career who try to find success with a proof of the Fermat problem. I gave some of the manuscripts to physicians who diagnosed heavy
schizophrenia.

I wonder in which class, nonsense or mathematics, the Akademie would have
put the work of R. B. Here is an excerpt from his 1983 proof of FLT, a sample from
page 23. The proof contained 33 pages in all.

Lemma 1
Consider

$$\sum_{i=1}^{R-t} \binom{R}{i} (2x^{cm}U_1)^{R-i} (2^cU_2)^i \qquad\qquad \text{I}$$

and

$$\sum_{i=t}^{R-1} \binom{R}{i} (2^cU_3)^{R-i} (2^{cm}U_4)^i \qquad\qquad \text{II}$$

where

$$\mathcal{P}U_1 = \mathcal{P}U_2 = \mathcal{P}U_3 = \mathcal{P}U_4 = 0 \qquad\qquad (65)$$

and $\mathcal{P}_a$ means par of sigma-term for $i = a$.

For any t in I,

$$\Delta_1 = \mathcal{P}_{R-t-1} - \mathcal{P}_{R-t} \quad [R > 2]$$

$$= \mathcal{P}\begin{pmatrix} R \\ t+1 \end{pmatrix} + cm(t+1) + c(R-t-1) - \mathcal{P}\begin{pmatrix} R \\ t \end{pmatrix} - cmt - c(R-t)$$

And for any t in II,

$$\Delta_2 = \mathcal{P}_{t+1} - \mathcal{P}_t$$

$$= \mathcal{P}\begin{pmatrix} R \\ t+1 \end{pmatrix} + c(R-t-1) + cm(t+1) - \mathcal{P}\begin{pmatrix} R \\ t \end{pmatrix} - c(R-t) - cmt$$

Thus, in either case,

$$\Delta \equiv \{cm - c + \mathcal{P}[(R-t)!] + \mathcal{P}(t!)\}$$

$$- \{\mathcal{P}[(R-t-1)!] + \mathcal{P}[(t+1)!]\} \equiv \mu_1 - \mu_2$$

$$\Rightarrow \Delta_{\min} = (\mu_1)_{\min} - (\mu_2)_{\max}$$

and

$$(\mu_1)_{\min} = m - 1 + \{\mathcal{P}[(R-t)!] + \mathcal{P}(t!)\}_{\min}$$

$$\equiv m - 1 + \xi_{\min} \quad [1 \le t \le R - 1]$$

$$\Rightarrow \xi_{\min} = 1 \Rightarrow (\mu_1)_{\min} = m$$

B. had mastered elements of the mathematical style, though it is not clear where the statement of the lemma stops and where its proof begins. The Akademie would have trouble with B.: his work certainly *looks* like mathematics, but it would take days or weeks to read all 33 pages of it. (Work like B.'s allows fermatists to make their claims that no one has ever been able to find an error in their proofs.) Probably it would be best to classify it as nonsense.

B. is not typical of fermatists, most of whom are unable to handle summation notation or the binomial expansion. More ordinary is S. A., who accomplished the proof of FLT in a 12-page pamphlet in 1972 by supposing, among other errors, that the negative of the statement

$$x^n + y^n = z^n \qquad \text{for all } n > 2$$

is

$$x^n + y^n \neq z^n \qquad \text{for all } n > 2,$$

which is not the case. A. circulated his work and got many negative responses, not one of which caused him to change his mind in the slightest. One mathematician, writing with the clarity and candor so often found among mathematicians, and with the acerbity that comes from dealing with cranks, replied in part:

> Unfortunately it would seem that your proof replaces accurate logic by a reliance on ambiguous use of the word "any". In short, your proof is no better than a bad pun. I don't imagine for a moment that you will accept this, although the point is really rather elementary and it should hardly be necessary to explain it in detail to you. It is really terribly sad when people waste their time in this way, but that is entirely your business. You really have two choices:
>
> (a) Buy a book on mathematical logic and understand it, in which case you will see at once why your proof is not only wrong, but idiotic.
>
> (b) Refuse to accept a judgment which any competent mathematician will give as soon as he looks at page 6 of your paper, namely that there is an obvious and serious error of a distinctly elementary nature. In this case, you will waste a great deal of your life in futile argument, attempting to deny a fact which is as plain as day. Which, if I may be frank, is the hallmark of a crank.

A., naturally, chose to take option (b), writing to me about the "most rude and patronizing letter" he had received. He also wrote:

> What you should do is to *prove* that an angle *cannot* be trisected, that a circle *cannot* be squared, that proof of FLT *cannot* be found. Unless you can do these things all you have is a distasteful feeling that these cannot be done, but no proof!

When I told him that the first two proofs had been done, his reply was:

> What I meant was that none of these proofs are intelligible to ordinary people! For instance, G. H. Hardy defines a "transcendental number" as "a number which is not algebraic"! Then he proves that e and π are transcendental, then he concludes from the latter that "circle cannot be squared"! Even people with degrees in Applied or Scientific or Engineering Mathematics would find these proofs difficult to understand, so what chance has any layman?

A. later abandoned FLT and produced *The Absolute Theory of the Universe*, a book of physics crankerly asserting, among other things, that the stars are not huge balls of inconceivably hot gases, but are in fact merely reflections of our sun. He somehow managed to get a full-page write-up in a newspaper, complete with a two-column picture of himself. The reporter wrote:

> He believes his lack of formal training may explain why people like Patrick Moore and Magnus Pyke have so far failed to acknowledge their review copies of his book.
> Mr. [A.] remains optimistic however, that this is a good sign.
> He says: "If they don't send any reply I think they can't find anything wrong with the theory."

Not likely.

The story concluded with:

> Mr. [A.] also claims to have managed to solve the baffling Rubik's Cube, which his son brought home recently.
> He says: "I was able to work out the mathematical theory of it."

Also not likely.

> "I managed to get one side done—but it seemed pointless to do the rest."

For "pointless" read "impossible." Why newspapers pay attention to cranks is hard to explain.

Nevertheless, they do. Here is a headline on page one—page one!—of a 1984 issue of the *Guardian*, one of the brightest jewels in the diadem of British journalism:

CODES AT RISK IN MATHS ADVANCE

There followed a long story on how A. A.'s proof of FLT (A. A. is not related to S. A.) would make it possible to

> find with ease prime numbers of any magnitude and to factor, very rapidly, large numbers that are themselves products of prime numbers.

This, of course, would compromise those cryptographic methods that depend on the non-factorability, in any practical length of time, of a 200-digit product of two 100-digit primes. Front-page news, indeed. However, there were three things not quite right. First, the prime number that A. gave as evidence of the power of his discoveries, $2^{214313833} - 1$, is divisible by $295,753,090,783$ and hence is not prime. Second, his proof made no sense. Third, he gave no connection between FLT and the problems of prime testing and factoring.

How did the *Guardian* get so taken in? I suppose it is the pressure to meet deadlines that keeps science writers from checking mathematical claims with mathematicians, and their lack of experience that keeps them from recognizing crankery when it is before them. Crankery should not be all that difficult to recognize, though. As one of the many responders to the article wrote:

> More distressing than the factual errors is the whole tone of the article, which displays numerous symptoms of crankiness: historical errors, vagueness and lack of clarity, an unduly simple answer to a complex problem, confusing notation, grandiose claims for the significance of the result, secrecy about some key ideas, private publication, persecution mania, rejection of standard mathematics, truisms presented as profundities, etc. I am dismayed that a journalist as experienced as [the reporter] should have been taken in by such rubbish.

I do not know what the "etc." could refer to; that splendid list of characteristics of cranks seems complete.

The *Guardian* had to take back its story. The reporter admitted that the alleged prime was composite, and that

> Many mathematicians have pointed out, with great vigour, that what we published of Mr. [A.]'s work did *not* offer a solution to Fermat's Last Theorem, and although like many humans, most journalists and even a few scientists, we cling wistfully to the thought that what we had believed to be true must have had *some* truth in it, we can hardly go on claiming to have printed a true solution. In our quantum world we find ourselves at the increasingly opaque end of a probability curve. We are therefore repentant.

Not quite as repentant as he should have been, considering the number of people who read the original article and probably never noticed the retraction. Some of them will carry away a feeling that FLT was proved, or that whether it was proved or not is a matter of controversy, or that the mathematical establishment was wrongly refusing to consider the work of an outsider. The reporter went on:

> We are not quite so repentant of the real point of the story—that techniques of encryption are at risk.

and he continued, mentioning dinosaurs and proton decay among other things. It was not the complete mea culpa that one might have wished for.

To its credit, the *Guardian* published a selection of letters on the subject under the heading

A Suitable Case for Embarrassment

though the last letter in the selection was by A. himself and concluded:

> A proper and responsible evaluation can only be made by a close reading of the full papers and by actually working the given maths.

The *New Statesman* took the opportunity to do some understated crowing by publishing the following note:

> The [A.] technique relies on persuading students and journalists that a smattering of nonsensical algebra and geometry reveals methods of breaking international military and diplomatic codes and the firing sequences for nuclear weapons. Earlier and prestigious victims of the [A.] technique have included the *Far East Economic Review*, *Computer Weekly* and minor science magazines. . . .
>
> A year ago, [A.] wrote to the *New Statesman* claiming that the Government Communications Headquarters, GCHQ—which is responsible for British codes and cyphers—was taking action against him to suppress his discoveries. He sent us his "unpublished" mathematical papers—the same ones the Guardian has now reported on. They were nonsense. Asked to substantiate his claim that GCHQ was suppressing his work, Mr [A.] withdrew. He asserted merely that GCHQ was so threatened by his discoveries that they would have to suppress his work in the future.

The article ended with a quote from a mathematician:

> The *Guardian*'s endorsement of this rubbish is dangerous in that it discredits serious comment on the applications of mathematics.

Exactly. The *Guardian* will no doubt be more cautious in the future. However, there are other newspapers and they all must be read with a critical eye.

S. A., the fermatist preceding A. A., wrote that

> You can find out a lot about cranks in general from the British American Scientific Research Association from the following address. (About half the members have PhD or above from reputable Universities).

Naturally, that had to be followed up. J. E., the secretary of the Association and editor of its quarterly *Journal*, replied:

> Of course, [S. A.] is correct in his statement. I know of no better collection of cranky material than that to be found in our Journal, which, however, concentrates on fundamental physics rather than mathematics. There are two pathognomonic signs of a crank: one is belief that he has proved Fermat's Last Theorem and the other is a disbelief in the sacrosanct

theory of special relativity. There is little doubt of the classification of Mr. [A.] himself. I myself am also to be classified as a crank on both grounds. Hence I have proved Fermat's Theorem and support such disbelievers in relativity as the late Professor Herbert Dingle and Professor Stefan Martinov, the dissident Bulgarian physicist, who was threatened with deportation from the U.S.A.

E. kindly sent me some material, including his proof of FLT, and a sample copy of the *Journal*. Its title page included

> BASRA has now completed twenty years of successful research and creative scientific innovation. There has been a constant and productive exchange of ideas between members, and many new concepts have been introduced into the different branches of knowledge. It would be misleading to suggest that all these concepts are sound and will survive as pillars of our main body of knowledge in future ages, but such an exchange of opinion and criticism must inevitably be creative. It is safe to say that the reputation of our association will be far greater over the next twenty years than it has been in the past.
>
> Neither lack of academic qualifications nor non-membership are barriers to publication in the journal, and advice regarding the preparation of copy may be obtained from the editor.

The issue included three articles (one by E.), an editorial, notices of eight publications (including A.'s *The Absolute Theory of the Universe*), and five letters. I did not subscribe to the *Journal* since the contents of that issue were free of mathematics. They were not free of crankhood, however.

That issue was published more than ten years ago. BASRA carries on today, E. still edits its *Journal*, and he still thinks that he has proved FLT. He wrote recently:

> Professor [S.]'s letter is indeed badly flawed. It appears to have been so structured as to avoid having to acknowledge any previous work by others, but it has given me the way to complete the proof.
>
> The proof centers around the factor $\sqrt[n]{-1}$ and therefore must be closely related to the proof offered by Gabriel Lamé (? at the French Academy of Sciences) and refuted by Joseph Liouville. I have not any easy access to this work, but it might be worthwhile to review it. I feel there was probably a similar error in it as there was in Euler's proof that the equation was impossible for $n = 3$. Once one is on the right track the proof is so simple that it will not require a panel of eminent mathematicians to investigate it.

E.'s proof, which is still undergoing revision, is similar in form to the proof of many fermatists: pages, in some cases a few and in other cases many, of elementary and correct reasoning followed by a quick descent into incomprehensibility, then a return to the surface with something like "therefore $x^n + y^n = z^n$ has no solutions in positive integers" that does not follow from what has gone before. There is no therefore there.

E. and his Association show that there exist cranks who are educated and literate (in case you do not know the meaning of *pathognomonic*, as I did not, it is "typical of a particular disease"—definitely the *mot juste*), and who do not reject the label of crank. Many people have the impression that cranks are ignorant, wild of eye, or both, and while that can be the case, it is not always so. Most people think that cranks do not know that they are cranks, but E. shows that it is possible to acknowledge the label with equanimity and good humor. He wrote:

> The study of this material [works of cranks] is a legitimate field of research. What is their basic psychology? Is it some form of ego defense? What types of error are they subject to? And, in what way does this correlate with the basic psychological state? However, in all fairness I believe such a study should be paralleled by one investigating the resistance of the scientific establishment to innovatory changes. The treatment of Einstein's special relativity is a specially good example of this. Whittaker in his classic history remarks that in spite of the brilliance of Newton's work, it was not readily or quickly accepted either in Britain or France.

The last part of that has echoes of the classic crank syllogism: "Galileo was persecuted, Darwin was vilified, Newton was not understood, all of them were right; I am ignored, vilified, and misunderstood, therefore I am right too."

It is a question whether publications like the *Journal* perform a service and are thus good, or promote confusion and are therefore bad. On the one hand, the number of cranky ideas in it that will later turn out to be valid is surely very close to zero. On the other hand, perhaps scientists need cranks and those convinced by them to keep them from becoming too smug and arrogant, and too convinced that they have the final say about truth. Which is worse, confusion or arrogance?

Another difficult question is, does paying attention to cranks do any good? E. P. proved FLT in 1986 and sent his proof to several universities. Its error, P.'s assumption that a square root had to be an integer when it did not, was easy to spot. A mathematician replied with a letter that was as long as the proof. Parts of it state universal truth and deserve repeating:

> As a professional mathematician, I should warn you that very few mathematicians will take the time to examine supposed proofs of this

conjecture. You should thus not expect to receive many replies to your inquiries. The fact is, some of the most brilliant mathematicians of the last few centuries have worked on the problem, so that we are supremely confident (on commonsense grounds) that no short proof exists, and that Fermat himself was undoubtedly mistaken in his belief that he had a proof. Most professional mathematicians don't want to waste their careers on such a tough nut, so that the number of well-trained mathematicians now looking at the problem is rather small. On the other hand there have been literally hundreds of thousands of amateurs who thought they had a proof—all convinced they were right, all completely mistaken. That is why most professionals won't even bother to answer letters like yours. All this is by way of preface to my main remarks, to which I now turn, trusting that you will forgive me for speaking with absolute candor.

To be perfectly frank, you have managed to crowd a very large number of mistakes into just three pages. Some of these mistakes are the result of faulty drafting and could be corrected by more careful writing. Others, however, are genuine mathematical mistakes and render your argument invalid.

[There followed three paragraphs on P.'s mathematical and logical errors and omissions.]

If you find these words discouraging, I regret that you have wasted your time. On the other hand I thought it kindest to tell you your mistakes rather bluntly, so that you can avoid wasting any more of your time or the time of busy mathematicians. If you find Fermat's conjecture interesting, by all means continue to think about it. Remember though that the chances of any amateur solving this problem are microscopically small. As a way of gaining fortune or fame it is far inferior to trying to win a state lottery.

Harsh words, you may think. But here is P.'s reply, in its entirety:

Thank you for your informative reply to my paper concerning "Fermat's Theorem".

Your time and consideration is appreciated.

Thanks again.

Did the harshness work? Was the mathematician's effort in reading the paper and writing the reply—two hours of work at least, and probably more—worth it? Maybe so.

But more likely not. S. S., who proved the theorem in 1982, was unlike ferma-tists whose proofs have their first error on the last page. S. made mistakes early, so I was able to write:

> I find two errors on page 2. The first is "Let z be the smallest number greater than y." You have no control over the value of z and there is no reason why it should be $y + 1$. Similarly, you may not "Let z^2 be ...$x + y - 1$."
>
> The second error occurs in "Since $z < S$ and since both divide $x^p + y^p$ they must have similar factors." That does not follow. 6 and 9 both divide 162, but have different prime factors.
>
> I hope you will give up the fruitless effort of trying to prove Fermat's conjecture by elementary means. I have seen more than fifty such efforts, all incorrect.

Let no one accuse me of giving encouragement to cranks. However, I thought I knew what would happen, and I was right. S.'s reply included:

> Re your comment on the assumptions $z = y + 1$ and $z^2 = x + y - 1$. You apparently did not read the paragraph carefully. I was not trying to control the value either of z or z^2. I was simply setting the maximum possible range between the two to show that even with that range z^2 could not be within the arena of analysis. ...
>
> Re your comment regarding similar factors: I admit giving an incor-rect reason on page 2 to jump to a correct conclusion on page 3. ...
>
> Neither of your criticisms is, I feel, substantive, and furthermore, I get the feeling that you did not get beyond page 2 of my manuscript.

His feeling was correct. I did not reply and S. got mad at me, but I was acting for his own good. Given enough negative reinforcement, he will give up mathematics, I thought. However, three years later he broke into print in the pages of the *Bulletin of the Dozenal Society of America*. For some people the attraction of mathematics is so great that, when one door is barred, they look for another that may be open.

Every once in a while a proof of FLT contains something new. *A Proof of Fermat's Last Theorem*, a 1980 manuscript by A. C., starts by replacing $(x+y)^n$ with its expansion in powers of x and y, a tip-off that its author knew some mathematics. In the course of the proof, he defined a sequence of polynomials by

$$s_0 = 2, \; s_1 = x, \ldots, \; s_{n+1} = x(s_n - s_{n-1}), \qquad n = 1, 2, \ldots,$$

so you get

$$s_2 = x(x - 2) = x^2 - 2x,$$
$$s_3 = x^3 - 3x^2,$$
$$s_4 = x^4 - 4x^3 + 2x^2,$$
$$s_5 = x^5 - 5x^4 + 5x^3,$$
$$s_6 = x^6 - 6x^5 + 9x^4 - 2x^3,$$

and so on.

C. then said that the sum of the coefficients of s_n is $2\cos(n\pi/3)$. He was right, and seeing why is a pleasant exercise. His polynomials may have other properties that could also provide pleasure. It looks as if every coefficient in the derivative of s_n is divisible by n. Maybe the fact that

$$x^2 = s_1 s_2 - s_3$$
$$x^3 = s_2 s_3 - s_5$$
$$x^4 = s_3 s_4 - s_7$$

has something to do with it. Those polynomials have possibilities.

Demonstracion de la Conjetura o "Ultimo Teorema" de Fermat, by E. L. of Colombia (1982, revised 1983), consists of six densely typed pages of an odd size: $8\frac{7}{16}''$ by $13\frac{13}{16}''$. Yes, I know Colombia uses the metric system, but 34.5 cm by 21.3 cm is also an odd size.

> Copias de esta demostracion han sido enviadas a: Universidad de Heidelberg, Universidad de Yale, Academia de Ciencias de Francia, Revista Episteme de Mexico, Academia Colombiana de Ciencias, Universidad de Oxford, Revista Episteme de Mexico, Professor Harold Edwards del Courant Institute de New York, Mr. George Seligman Universidad de Yale, Barry Mazur Universidad de Harvard, Universidad de Goetingen, U. de Wupertal.

"Copies of this demonstration have been sent to . . . ." The list causes one to reflect on the advantages of obscurity. My Spanish was not up to reading the proof, even though it looked elementary.

The reply to L. by the French Academy of Sciences referred to the decision made by that body on May 28, 1947 not to look at any more proofs of Fermat's Last Theorem. Part of it follows:

Les personnes qui s'appliquent à donner la démonstration attendue, ignorent généralemant le point d'avancement de la question ou commettent des erreurs de raisonnement assez souvent renouvelées. C'est une grand et inutile perte de temps pour les Membres de l'Académie d'examiner les pretendues demonstrations qui parviennent fréquemment à la Compagnie.

"It is a great and useless waste of time for the Members of the Academy to examine the alleged demonstrations that frequently come to the Company." Gallic precision!

And Gallic wisdom as well. Now I too am wise and know that it is in general not useful to correspond with cranks, but I was not always so. Wisdom comes mostly from experience and every generation repeats the errors of the one before, even though every generation has been told, very clearly, not to make them. The following correspondence illustrates some of the experience.

February 25: 21-page handwritten manuscript arrives from G. M., proving FLT. February 29: letter to M. asking for clarification of a point. March 6: two-page reply from M., brimming with gratitude. March 27: letter to M. with the blunt statement

Theorem 1 on page 3 of your manuscript is incorrect,

followed by worked-out counterexample showing that the value of (2, 441) is 42, while M.'s Theorem 1 would imply that it is 126. (What M. meant by (2, 441) is irrelevant, both to this discussion and to FLT.) April 9: M. replies with three new pages, all correct, to replace the one old page with the error. May 25: after no answer from me (it is always easy to find something better to do, or that needs doing more, than looking at a 23-page proof of FLT), another letter, ending:

Undoubtedly, your own generous participation in this matter has already extended quite beyond my original anticipation. I continue to be appreciative.

May 25: (this letter to M. crossed his letter of May 25 in the mail) on page 12, a congruence implied either (a) or (b):

In your work, you assume that (a) must hold. It seems that you must also consider (b) and show that it too leads to a contradiction.

June 3: M. countered with:

The enclosed theorem, derivation and examples (sheets 22–26) are directed toward resolving this issue, as well as some others which caused me slight uneasiness.

And there it ended. I forget if I could not understand the revision or if I even looked at it, but that was the end of M.'s proof for me. Was anyone the gainer?

More evidence of the inutility of convincing fermatists that they are wrong, if more is needed, can be found in the 1976 work of R. P., *Fermat's Last Theorem: A Remarkable Proof*, "published by Theorem Publishing Company." It contained a proof that could not be refuted because it did not have any statement that anyone could point to and say, "That is wrong, for this reason." It did have many statements to which anyone could point and say, "I don't understand that. What does it mean?"

P. found a patient mathematician to correspond with, one who wrote at least two two-page letters, pointing out errors and making statements like

> Finally, a leap of logic you perform in the final two sentences of page 4 continues to elude me.

Of course it did no good. Nor did the mathematician's closing sentence in one letter:

> In summary, I found your monograph refreshing, interesting, and thought-provoking. I hope you can resolve the above difficulties.

Encouraging cranks is folly. Responses to the objections flowed freely, and the correspondence ended with P.'s grumpy final letter:

> Your objection to my proof of Fermat's Last Theorem has been received and noted.
>
> May I refer you to the last line of the Preface: And yet it exists, for those with the perception to see it.

That is how it usually ends. Except for the crank it does not end. I have another copy of P.'s proof, this one dated 1984. It had grown to 18 pages since 1976, but with no increase in clarity. P.'s letter to me contained:

> Has not the same advanced mathematics that Fermat was capable of been available to at least some of the mathematicians who have since worked on the problem? If this were not so, would there not be other theorems of his that would still be unsettled? Why, then, is this problem still unresolved? There is something apart from advanced mathematics that has not been seen, that has not been taken into account.

I am not sure I follow that, either. There are some people who look perfectly ordinary, but whose minds work in strange and unfathomable ways.

L. A. constructed *A Proof of Fermat's Last Theorem* in 1982. He had heard about the Wolfskehl Prize:

> I understand there are prizes for the correct proof, but I have no idea who to contact to receive them. Perhaps you can help me.

Here is his proof. See if you can follow the logic. Suppose that

$$x^2 + y^2 = z^2 \quad \text{and} \quad x^n + y^n = z^n.$$

Now let

$$V = x^2 + y^2 \quad \text{and} \quad W = x^n + y^n.$$

When $x = y = 1$, it follows that $2 = V = W$. From $V = W$ we get, from the previous equation,

$$x^2 + y^2 = x^n + y^n.$$

From that follows, from the first equation,

$$z^2 = z^n.$$

Since $n > 2$ and $z > 0$, that can be true only for $z = 0$ or $z = 1$. But that is impossible, since

$$z^2 = z^n = V = W = 2.$$

It was no surprise to me that A. failed to appreciate the reasons I gave him why his proof was not correct.

R. B., the fermatist whose lemma was quoted before, began working on FLT in 1951 and is still at it. He wrote:

> I'm no mathematician; but I love it, what little I can dabble in.

Even though no mathematician, he knew that Mertens' conjecture had recently been disproved. There are many mathematicians who could not state Mertens' conjecture.

If anyone had worked through B.'s 33-page proof, the work would have been wasted because B. later discovered that it was not perfect:

> Three months ago I found (not an error, but) an insufficiency in my work on Fermat's Last Theorem. . . .
> Then I conceived a new gambit. . . . I think now this has resulted in proof of F.L.T. for all exponents (above some finite value not yet fixed— probably 6000 or so) and of the form $1 + 4f$ ($f > 0$). This leaves the other class, $m = -1 + 4f$, which, so far, is resisting my efforts.

B. could recognize error, so he might better be classed as a mathematical amateur rather than as a crank. But maybe not: besides the proof of FLT, he sent me a proof of the Goldbach conjecture that contained an obvious logical error. When I pointed it out, he refused to acknowledge the validity of my criticism. Perhaps crank is the more accurate description. He was, however, a *pleasant* crank, unlike many who are bitter, enraged, hostile, or all three.

F. B. provided evidence that the Fermat bug can bite anybody. B., who wrote in 1974, was a woman and had a PhD in mathematics, two characteristics that tend to provide immunity to mathematical crankery. If we let W denote the class of women, D the class of holders of PhD's mathematics, and F the class of fermatists, then $W \cap D$ already has small cardinality, $W \cap F$ is smaller still, and $D \cap F$ is certainly not much larger. I would conjecture that the number of elements in the set $W \cap D \cap F$ (which is a variable, a function of time) is bounded above by 2 for any time t.

B. wrote:

> During my forty seven and one half years of teaching, I taught classes and gave students extra help by day, while at night I graded papers and prepared lessons. So there was no time for research, nor time for research done by others. Moreover, I did not have ready access to a university library. So I knew very little of what has been done on Fermat's Last Theorem.
>
> In [19xx] I was granted a Ph.D. in mathematics by the University of [O.]. Professor [C.], under whom I wrote my dissertation, said it was too long and contained too much computation to be sent to a journal for publication. So I have never had anything published.
>
> In September of [19yy], I sent what I hoped was a proof to Professor [S.]. He followed my reasoning up to the last page, and then he thought I had failed to prove anything. Also, Dr. [R.] of the [W.] College faculty found two counterexamples showing that what I thought I had proved is not true in all cases. I hope that this attempt was successful.

It was not: there was an error on page 4. B. replied to that finding:

> It appears that my arguments on pages 4 and 5 need further clarification.

She provided what she hoped was clarification, but the person to whom she wrote (not me) answered:

> Your reasoning on p. 5, line -7 is therefore utterly and totally specious. There is no justification for the "Therefore".

That was probably enough to discourage her forever.

Another attempt at FLT by a woman was made in *An Attempted Proof of Fermat's Last Theorem by a New Method*, by C. W. This was a real book, though paperbound and only 41 pages long, and was published in 1932 by G. E. Stechert and Company. Stechert also published Bolza's *Lectures on the Calculus of Variations*, Boole's *Treatise on the Calculus of Finite Differences*, Gow's *A Short History of Greek Mathematics*, Whitworth's *Choice and Chance*, and many other well-known books, some of which are still in print. W. wrote a modest preface, which started:

> I publish this little work by itself in order to invite criticism. The argument employed seems to me flawless, but others may not find it so. If the final judgment is adverse, no harm will be done, and it may possibly lead to something better.

Here is an outline of part of W.'s proof. Start as usual with

$$x^n + y^n = z^n,$$

with n an odd integer. Transpose y^n, and factor:

$$(z^{n/2} + y^{n/2})(z^{n/2} - y^{n/2}) = x^n.$$

Now put

$$z^{n/2} + y^{n/2} = p \qquad \text{and} \qquad z^{n/2} - y^{n/2} = q,$$

so

$$x^n = pq, \qquad y^n = \left(\frac{p-q}{2}\right)^2, \qquad \text{and} \qquad z^n = \left(\frac{p+q}{2}\right)^2.$$

Nothing wrong so far, since nothing has been said about anything having to be an integer. Now, W. said, suppose that (x, y, z) is a nontrivial solution in integers. There are three cases: either p and q are both rational, one is rational and one is irrational, or both are irrational. W. said that in the first case, the last equations show that p and q must be integers. Thus y and z raised to an odd power give perfect squares, so they must themselves be perfect squares. So, put $y = s^2$ and $z = r^2$ to get

$$r^{2n} - s^{2n} = x^n,$$

or

$$(r^n - s^n)(r^n + s^n) = x^n.$$

Then, since under the usual assumption that x, y, and z are relatively prime it can be shown that the two factors on the left-hand side of the last equation are relatively prime, both factors must be nth powers. Thus,

$$r^n + s^n = t^n \qquad \text{and} \qquad r^n - s^n = u^n.$$

The first equation is another solution of $x^n + y^n = z^n$ in smaller integers than the solution we started with. So, by Fermat's method of infinite descent, it follows that there are no nontrivial solutions.

W. commented:

> Is it possible that Fermat used this argument to prove his theorem, and at first was satisfied with it, and hastily wrote in his copy of *Diophantus* that he had a proof of it—"wonderful" because simple and tolerably short, though too long to insert? Then later he may have noticed what was lacking in his proof, and not being able to use further the argument by descent, to which he was wedded, he abandoned his theorem, never again alluding to it. If, however, he did supply what was lacking, there is one more possible (though not very probable) explanation of his silence about the theorem: it may be that he discovered it only late in life, and therefore had no occasion to call the attention of others to it.

I agree with W.'s reading of history. However, there are two cases to go in the proof. To have one of p and q rational and one irrational is clearly impossible, for then x would be irrational. Thus we are left with both p and q irrational, a case that W. treated briefly on pages 20 and 21 of her book. The argument was neither clear nor convincing to me, nor evidently to anyone else at the time the book was published.

Das Fermat'sche Problem, by H. D., published in 1933, is a nicely printed 18-page pamphlet, full of equations and even a few congruences modulo $2^{\lambda+1}$. D. also had heard of the work of Lamé. The details of his proof were too much for me, but D. is notable because he may provide evidence that crankhood is not incurable. The person who kindly sent his work to me wrote:

> [D.] was a friend of my grandfather's who, having neither knowledge nor interest of the subject, greatly appreciated his friend's mathematical achievement. [D.] was a civil engineer, a qualification that required little mathematical education at that time, and he had attained a position of modest prominence in the civil service of Bavaria. But he was worth more than his job would signify. 1933, when he published his solution of this long unsettled problem, he stood in the end of his fifties. The subsequent turmoil of the times helped him to come to terms with the fact that his contemporaries were unable to value his contributions to science. When

he died in the 1950s, he was apparently by no means a crank (any more?) but a kind old gentleman.

R. D., a retired colonel, mailed his proof of FLT to more than 125 mathematicians. His prose was quite unlike what you might think a colonel would write. However, that is not altogether surprising, since the same sort of thing can be seen among some mathematicians. When writing mathematics, mathematicians produce tightly controlled, disciplined, and logical prose. Some of them, when writing outside of mathematics, produce prose that is violent and emotional. It is as if the requirements of the discipline can control something under pressure, but when they are taken away there is an explosion. When the discipline of the army was removed D. may have reacted similarly. Colonels cannot be wild men, can they?

In his proof, D. started with $x^n + y^n = z^n$, replaced z with $K + H$ and x with $K - H$, let $y = f(K, H)$, and then came the

P∗R∗O∗O∗F! *This must-important procedure*, immediately above, *generates at least once all potentially-admissible values of x, y, z*— resulting in the following (in passing it would be of interest to see that *one such admissible set of x, y, z* for a given *n must admit of an infinity of such* (different) *sets* for the same *n*):

$$(K + H)^n - (K - H)^n = f(K, H)^n—$$ (1)

an *IDENTITY*!

THE GUTS! This means in one simple transformation our problem changes—metamorphoses!—*from the equation demanding the endless examination of an infinity of number-set-solutions (potential!) to the IDENTITY involving a GREATLY-SIMPLIFIED REVIEW of the relatively very-few potentially-applicable "function-forms," to determine if any will satisfy the IDENTITY as to the replacement of y, f(K, H),*
(WHEW!)

In general this function-form examination is very economical of taut-time and mathematical-maturity—not to mention elimination of *vex-hexing dullness* incident to the monotony of endless applications-in-iterations of the same technique(s) in attempting the equation-solution.

This is well. *"Ars est longis, vita brevis!"*

Fermatists must live in a state of tension. On the one hand, they want their work to be accepted so they want it to be clear; but on the other hand they do not want anyone to find any errors in it, so they do not want it to be *too* clear. This may explain why fermatists' proofs are hard to follow. There is also the possibility that they are writing with as much clarity as they can muster, and this may apply to D.'s proof.

The first explanation is more applicable to *Congruence Surds and Fermat's Last Theorem*, an actual hardcover book by M. M., published in 1977. It was published by a vanity press, of course, but perhaps M. got his money's worth by being able to handle and admire it and read the copy on its dust jacket, no doubt written by himself:

> The relentless but unsuccessful search by the mathematical profession for the rediscovery of Fermat's "marvelous" proof of his Last Theorem has gone on now for 300 years. The simplicity of the rediscovered proof is commensurate with the simplicity of the theorem. But the proof is very far from obvious.
>
> It appears that the mathematicians looked up too high, higher and higher, instead of looking down at the humble digital writing of numbers. Such writing is so well understood, everything about it has already been said. So it seems. But not so. There is never a limit to more clarity and to even more profound understanding.

M., a retired professor of mechanical engineering and aeronautics, had either discovered or been very impressed with how division could go from left to right instead of in the usual way. To divide 1995 by 7 we can write

$$
\begin{array}{r}
7\ )\ 1995\ (\ 285 \\
\underline{14} \\
59 \\
\underline{56} \\
35 \\
\underline{35}
\end{array}
$$

as we were taught in school, choosing the digits in the quotient from left to right by seeing how many times 7 goes into what we have left. Or we can go the other way, choosing the digits in the quotient from right to left so as to make the rightmost remainder zero at each step:

$$
\begin{array}{r}
7\ )\ 1995\ (\ 285 \\
\underline{35} \\
96 \\
\underline{56} \\
14 \\
\underline{14}
\end{array}
$$

First the 5, then the 8, then the 2, and everything comes out even. That is fun, and you might wonder why the method is not taught in the schools. The reason is that it is harder than the ordinary process when the divisor is something other than 7. For example, in the problem

$$5 \;) \; 1995 \; ($$

choosing 1, 3, 5, 7, or 9 for the rightmost digit in the quotient will make the rightmost remainder zero, but unless you make the proper choice you will not be able to proceed. Three will not do:

$$
\begin{array}{r}
5 \;) \; 1995 \; (\quad 3 \\
\underline{15} \\
98
\end{array}
$$

Stuck! Nine is the only number that will work, then another nine, then a three:

$$
\begin{array}{r}
5 \;) \; 1995 \; (\; 399 \\
\underline{45} \\
95 \\
\underline{45} \\
15 \\
\underline{15}
\end{array}
$$

Decimals come out differently when the right-to-left system is used. For example, when we divide 2 by 3 we get

$$
\begin{array}{r}
3 \;) \ldots 000002 \; (\ldots 3334 \\
\underline{12} \\
\ldots 99999 \\
\underline{9} \\
\ldots 9999 \\
\underline{9} \\
\ldots 999 \\
\underline{9} \\
\ldots
\end{array}
$$

Thus, $2/3 = \ldots 3334$. We check division by multiplication and, sure enough, $2/3$ times 3 is 2:

$$
\begin{array}{r}
\ldots 3334 \\
\times \quad 3 \\
\hline
\ldots 0002
\end{array}
$$

This representation of numbers may strike a chord, but M. said it is not the right one:

> The congruence surds in the above Fermat proof have been mistaken for p-adic numbers by all mathematicians who read the proof. This misunderstanding seems to be the foremost obstacle to understanding the proof. Surds are not p-adic numbers.

M. had given p-adic numbers a try, but they were too much for him:

> In 1973, Kurt Mahler published his notes on p-adic numbers under the title of an introduction. They are not that but require one. His notes are written in such super-technical language and symbolics that nobody in need of an introduction can understand them.

M. concluded that the lack was not in him, but in others. About Hensel, the originator of p-adic numbers, he wrote:

> Strange as it sounds, Hensel himself was also not clear regarding other aspects of his p-adic numbers, and his lack of clarity seems to be contagious.

Speaking of lack of clarity, here is the start of M.'s proof of FLT:

Fermat's integer equation is

$$A' = C' - B', \quad C' = c'^p, \quad B' = b'^p \tag{1}$$

Divide by $Q = q^p$, q being a large enough prime number dissonant modulo p.

$$A = C - B, \quad C = c^p, \quad B = b^p \tag{2}$$

Form the compound $(-B).C$, $-B$ being written as a purely periodic surd and C as purely periodic decimal number, both being joined in a compound.

$$A = (-B).C \tag{3}$$

If $A = 7 - 4$, what you do is write -4 as a surd, which is easy to do since -4 is that which when added to 4 gives zero:

$$
\begin{array}{r}
\dots\ 99996 \\
+\quad\quad 4 \\
\hline
\dots\ 00000
\end{array}
$$

Thus $7 - 4 = \ldots$ what? $7 + (-4) = 7 + \ldots 99996 = \ldots 00003$, which does not seem to lead anywhere.

M. continued, without further explanation:

> We discussed before exact-power compounds in which one of the two constituents has an exact-power dissonant period. But now both constituents have such period. Therefore both root periods are restricted, the rational root decimal is restricted to C, and the rational surd period is restricted to B. (The root may of course be non-periodic.) The only rational root permitted is therefore $(-B).C$. But this is excluded for an entirely different reason. It represents the fraction $C - B$, and it can never be equal to the correct root.
>
> It follows that the root of (3) can never be rational. Multiplying it by Q gives A' and this therefore also can never be rational. Fermat's Last Theorem stands thus proven.

What root of (3)? Nothing that has gone before in his book prepares for this. My conclusion was that M. wanted to have a proof of which he could say, "No one has ever been able to point to anything that is wrong in it." Statements such as "That's nonsense" would not count with M. because they do not refer to anything *specific*. I suspect that the final form of M.'s proof was the result of deletions of things to which objection had been raised.

M.'s book has one value. He noted that

$$\left(25\frac{2}{5}\right)^3 = 16{,}387\frac{8}{125}.$$

This raises the interesting question, exactly when does it happen that

$$\left(a + \frac{b}{c}\right)^3 = n + \frac{b^3}{c^3}$$

for positive integers a, b, c, and n with $b < c$ and $(b, c) = 1$?

A very common error of fermatists is to suppose that, because two expressions are not algebraically identical, they can never be arithmetically identical. They note that $x^2 + 1$ is a different polynomial from $x + 1$ and conclude that they can never be equal, which is not the case as taking $x = 1$ shows.

J. R. illustrates the error very clearly. The clarity probably results from his scientific training: the resume he included with his proof gives his education as

University of [X.], B.S. Chemistry, 1966
[X.] State University, M.S. Physics, 1972
University of [Y.] School of Medicine, M.D., 1976

His job history shows a career that for some reason went off the rails:

1966–1969: Math-science teacher for the U.S. Peace Corps
1969–1976: Graduate work in physics and medicine
1976–1979: Postgraduate medical training and general practice
1979–1988: English teacher and work for temporary employment contrac-
tors doing office and light industrial work

In his proof of FLT, he let

$$x = J + s, \qquad y = J + a, \qquad \text{and} \qquad z = J + b,$$

so $x^n + y^n = z^n$ becomes

$$(J + s)^n + (J + a)^n = (J + b)^n,$$

or, after using the binomial expansion and rearranging terms,

$$J^n + ns J^{n-1} + \binom{n}{2} s^2 J^{n-2} + \binom{n}{3} s^3 J^{n-3} + \cdots + s^n =$$
$$n(b - a) J^{n-1} + \binom{n}{2}(b^2 - a^2) J^{n-2} + \binom{n}{3}(b^3 - a^3) J^{n-3} + \cdots + (b^n - a^n).$$

Now, R. said,

> Since J, s, a and b are all rational number variables, one can look
> for solutions to the above equation by equating coefficients of J.

One *can*, but one does not *have to*. Equating coefficients gives, R. said,

$$s = b - a$$
$$s^2 = b^2 - a^2$$
$$s^3 = b^3 - a^3$$
$$\vdots$$
$$J^n + s^n = b^n - a^n$$

Since there are three quantities trying to satisfy n equations, it was not hard for R. to show that the equations have no solutions in positive integers (the first two alone give $a = 0$), and he concluded that

> Hence, there are no rational or whole number solutions for Fermat's
> equation for $n = 3$ or more, and the theorem is proven.

The conclusion does not follow. The same argument applied to

$$x^2 + 1 = x + 1$$

shows that, by equating coefficients of powers of x,

$$x^2 = 0$$

$$x = 0$$

$$1 = 1$$

and hence the only possible solution is $x = 0$. The arithmetical solution $x = 1$ is not an algebraic one. Fermatists find it hard to understand that difference.

A. F., who wrote *A Proposed Proof of Fermat's Last Theorem*, may be a representative of a class: people who think they have proved FLT, made a mistake, and who then drop the subject when it is pointed out. It is impossible to know how large this class is because people seldom write letters saying "Thank you so much for showing that I was mistaken."

When F. considered the third-degree equation, he wrote it in the form

$$m^3 + n^3 = (n + u)^3,$$

which is perfectly all right, but then he said

Let $m^3 = x^2$, $n^3 = y^2$, and $(n + u)^3 = (y + 1)^2$.

"You can't," the variables protest, "you can't make us do that." When this was pointed out to F., nothing further was heard from him. He had the choice of giving up or becoming a fermatist, and he probably made the wise decision.

More examples of fermatists could be given since there is no end to them. Even if FLT is proved, fermatists will no doubt continue to offer simpler proofs, as has happened with the four-color theorem. The best thing that could happen as far as dealing with fermatists is concerned would be the discovery of a counterexample to the theorem, so cards could be printed up saying

Your proof of FLT is incorrect, since $x^n + y^n = z^n$ for $x = 31415926535$, $y = 27182818459$, $z = 31431080767$, and $n = 58598744994$.

However, that will not happen, since it is known that if a counterexample exists it involves numbers much larger than those. Fermatists will never be put off that way.

Nor can much of anything else be done to discourage them. FLT is firmly embedded in the popular mathematical culture, so it will always attract amateurs who think that proving it is only a problem in algebra and that the failure of mathematicians to find a proof can be overcome by a lucky stroke on their part. It would be a good idea in 2007 to publicize widely the expiration of the Wolfskehl prize, which would eliminate the fermatists who were in it for the money. However, there would probably still be many left for whom glory would be a sufficient reward.

FERMAT'S LITTLE THEOREM

The name "Fermat's Little Theorem" is not satisfactory since the theorem is not little at all and its initials are FLT, but the two theorems to which Fermat's name is attached have to be distinguished in some way. The little theorem says that

If p is prime and a is not a multiple of p, then a^{p-1} is a multiple of p.

Sure enough: if $p = 11$ and $a = 5$,

$$5^{10} - 1 = 9765624 = 11 \cdot 887784.$$

For $p = 17$,

$$5^{16} - 1 = 2^6 \cdot 3 \cdot 13 \cdot 17 \cdot 313 \cdot 11489,$$

and so on for every prime. Of course, theorems are not proved by looking at examples; the proof in all texts on elementary number theory is quick and elegant and requires no calculations at all.

The theorem was discovered because of curiosity about perfect numbers, like 6, 28, and 496, that equal the sums of their proper divisors. The ancient Greeks knew, and Euclid wrote in his *Elements*, that a number is perfect if it can be written in the form $2^{p-1}(2^p - 1)$, where p and $2^p - 1$ are both prime. The first three perfect numbers fit this form with $p = 2, 3$, and 5, respectively. To find perfect numbers, we need numbers $2^p - 1$ that are prime, and we need to know when numbers $2^n - 1$ have divisors. Thus, we look at numbers $a^n - 1$ and discover Fermat's Little Theorem. The theorem has many uses.

Most mathematical cranks and people with the potential to become mathematical cranks have never heard of the theorem, so it is no surprise that it does not occur often in the literature of crankery. It may be a surprise that it occurs at all, since it is not easy to imagine how anything cranky could be said about it: it is a theorem, a nice theorem, proved once and for all, and there is nothing more to be said. Not so. With every theorem comes its converse, and the converse to Fermat's Little Theorem with $a = 2$ is

> If $2^{p-1} - 1$ is divisible by p, then p is prime.

The truth of a theorem does not imply the truth of its converse, and the converse (with $a = 2$) is false: $2^{340} - 1$ is divisible by 341, but $341 = 31 \cdot 11$ is not prime.

A mathematical amateur rediscovered the theorem with $a = 2$. That is, he noticed that $2^{p-1} - 1$ is divisible by p when p is prime for some small values of p, but he made no attempt to prove it. He also noticed that $2^{n-1} - 1$ is not divisible by n when n is 4, 6, 8, 9, 10, or 12. He then jumped to the conclusion that what is true for small values of n must be true for all values of n, and wrote to a mathematician about his two discoveries. He was duly congratulated for his rediscovery of a special case of Fermat's Little Theorem, and duly informed that $n = 341$ gave a counterexample to his second discovery. For ordinary people, that would be the end of it: after an "Oh (pause) I see" we would go about our business. Not the rediscoverer, though. He wrote back, concerning 341,

> But the number you arrived at is the *product* of primes.

It may have been a pro forma objection, a final twitch before subsiding. I hope so.

FIFTH POSTULATE, EUCLID'S

Why is geometry taught in school? Even though the experience of geometry is almost universal in the United States, not many people think much about it after they leave their high-school geometry classes. I expect that the average citizen's response to a question about learning geometry would be along the lines of, "Yes, well there are triangles and circles everywhere and everyone needs to learn about them, and proofs make you think logically. But I was never very good at them." What everyone learns about triangles and circles is soon forgotten, except that π has something to do with circles, and there has never been any evidence that proving an angle inscribed in a semicircle must be a right angle has ever made anyone think logically. Would not everyone be better off if geometry was replaced in schools by chess, equally good at inspiring logical thought and more likely to be useful after school days are over?

No. We need to study geometry. Geometry has made the human race what it is, and geometry deserves homage. When the semi-legendary Thales had the semi-legendary thought in the far-off sixth century B.C. that it was possible to prove that the base angles of an isosceles triangle were equal, deducing it from simpler things that were known to be true, he changed the course of human history in a big, big way. It was not one of the trivial turning points in history. What if Columbus's ships had all sunk in a storm? What if Napoleon had won the battle of Waterloo? What if Hitler had died in childbirth? Those are minor questions, inconsequential compared with "What if Thales hadn't invented deductive geometry?" Without Thales, there would have been no western civilization. Western civilization is based on science and reason, and without the example of geometry to show how powerful reasoning could be, rationality would never have caught on, science would never have developed, and the whole world would still be ruled by pharaohs and other god-kings. Reason

137

has been a good thing, and geometry was the first large success of reason. Without geometry, civilization would have had a much harder time existing, and it might have died. Praise to geometry!

Praise also to Euclid, who put geometry into a form so near to perfection that it went unchanged for more than a thousand years. Euclid's *Elements*,

> collecting many of Eudoxus' theorems, perfecting many of Theaetetus', and also bringing to irrefragable demonstration the things which were only somewhat loosely proved by his predecessors

(as Proclus wrote), was so well done that it became *the* model of reasoning. The *Elements* was used as a text for two thousand years. Two thousand years! No other non-religious book has been used for instruction, unchanged, for that long. It is a monument in human history. It is typical of human history that nothing is known about Euclid except that he lived in Alexandria while the first Ptolemy ruled Egypt. The dates of Ptolemy's reign are known exactly—306 to 283 B.C.—but Euclid's dates will be forever unknown. We know the name of Alexander the Great's *horse*, but not the name of Euclid's parents or children. In the history books, Alexander is "The Great"; Euclid is not "The Supremely Magnificent" or "The Unparalleled," he is just plain Euclid. Thus does history dispense its rewards: with no more justice than does society.

Euclid's *Elements* had great influence. When Isaac Newton finally came to write down calculus, he wrote it in the style that Euclid used. When Blaise Pascal's sister wanted to brag about her brother's genius, she said that the youthful Blaise had found for himself the theorems of geometry, *in the same order* as they appeared in the *Elements*, as though Euclid was expressing natural law that could be expressed in no other way. When Immanuel Kant wanted examples of synthetic a priori statements— truths about the world found by sheer reasoning—he found them in Euclidean geometry. The *Elements* inspired something close to worship.

The book was a model of how logic can work. It started with definitions of geometric terms. Some of Euclid's definitions were not what we find satisfactory today—instead of saying that a straight line is "a line that lies evenly with the points on itself," we prefer to leave the term undefined—but others were just fine:

> When a straight line set up on a straight line makes the adjacent angles equal to one another, each of the equal angles is *right*, and the straight line standing on the other is called a *perpendicular* to that on which it stands.

Now we know what a right angle is, and Figure 15 gives a picture of two of them. The purpose of the definitions is to let us know what we are talking about. There is no more logical place than that to start. After the definitions, Euclid included the common notions, things that are clearly true but are not restricted to geometry: two things equal to the same thing are equal to each other, equals added to equals are equal, things like that. No one can disagree with any of them.

Then there came the postulates, statements about geometrical objects from which, using the definitions, common notions, logic, and nothing else, all geometrical truth would be deduced. Part of the reason that the *Elements* is so impressive is that there are only five postulates. They are:

1. Two points determine a straight line.
2. Straight lines may be as long as you like.
3. Given a center and radius, a unique circle is determined.
4. All right angles are equal.
5. The fifth postulate.

The statement of the fifth postulate, translated from Euclid rather than paraphrased as the first four are, is

5. That, if a straight line falling on two straight lines make the interior angles on the same side less than two right angles, the two straight lines, if produced indefinitely, meet on that side on which are the angles less than the two right angles.

Postulates 1, 2, and 3 set out the rules of geometry: you may use a straightedge, it may be as long as you want it to be, and you may use a compass. No ancient Greek would maintain that they were anything but reasonable. Nor would anyone disagree with the reasonableness of the last two, even though they are different. Postulate 4 may seem not worth stating, but it makes the definition of an acute angle—one that is less than a right angle—meaningful. Without postulate 4, we might have two pairs of right angles with an angle α in the first pair that looks like an acute angle, since it is less than the right angle BOC (Figure 16). But is it really acute? Perhaps if it were compared with the right angle $B'O'C'$ it would turn out to be *bigger* than that right angle and thus not acute. Without postulate 4 there would be no way of knowing if an angle really was acute or not. Postulate 4 essentially says that geometric space is uniform: that no matter where we go, or how far away, right angles do not change. It is a tribute to Euclid's power of mind that he saw the necessity for postulate 4.

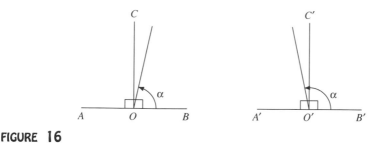

FIGURE 16

Postulate 5 is clearly different from the rest, if only because of the number of words that it takes to state it. It is like postulate 4 in asserting something about the behavior of geometric objects that can be far, far away. Given two lines cutting a third (Figure 17) where the sum of the interior angles is just a little bit less than two right angles—so little that the two lines could not possibly meet until they had gone more than three billion light years to the right—can we be really sure that they will meet? If you are like most people, who believe with all the strength of their intuition that Euclidean geometry is the only possible geometry, you will say "Of *course* they will meet. It stands to reason. The lines get closer and closer together and sooner or later they have to cross." It is so obvious that many people have tried to show that the fifth postulate is really a theorem, and can be deduced from the rest of Euclid's geometry.

FIGURE 17

Don't try to prove the fifth postulate. Euclid knew what he was doing when he made it a postulate. Consider postulate 4, that all right angles are equal. That stands to reason also, and is obvious. Maybe it is really a theorem, and can be proved. Yes! I have found it! Here is a proof that all right angles are equal. Consider two right angles, call them α and β (Figure 18). If the two are not equal, then one must be smaller than the other. Suppose that $\alpha < \beta$. Copy angle α twice onto the right-hand diagram, once on the right of the vertical line and once on the left (Figure 19). From the right-hand side of the diagram we see that

$$\alpha + \theta = \beta. \tag{1}$$

FIGURE 18

On the other hand, looking at the straight angle we see that

$$\beta + \theta + \alpha = \text{one straight angle.}$$

However, one straight angle is two right angles and α is a right angle, so, subtracting α from both sides of the last equation we get

$$\beta + \theta = \text{one right angle.}$$

But α is a right angle, so the last equation is

$$\beta + \theta = \alpha,$$

or

$$\alpha - \theta = \beta. \tag{2}$$

Adding (1) and (2) gives $2\alpha = 2\beta$ and hence the inescapable conclusion that $\alpha = \beta$. That is, all right angles are equal.

A convincing proof—what could be wrong with it? There are, in fact, several things wrong with it; it has hidden assumptions, hidden well enough to fool many people, but it would not have fooled Euclid. Euclid knew that it was impossible to prove that all right angles were equal, so he made it a postulate.

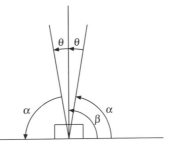

FIGURE 19

So it has been with the fifth postulate. Through the years many people have tried to prove it, but all their proofs rested on some more or less hidden assumption. If the hidden assumption were true, then the fifth postulate would be true. Thus, they replaced the assumption of the fifth postulate with a different assumption. That may be worth doing, but it does not prove that the fifth postulate follows from the other postulates.

The first attempt that we know of to prove the fifth postulate was made by Posidonius in the first century B.C. He did it by using a definition of parallel lines that was not the same as Euclid's, but that he must have thought amounted to the same thing. Euclid defined two lines to be parallel if they never meet. Posidonius changed that by saying that two lines are parallel if the distance between them is always the same. With that definition of parallel, he was able to show that the fifth postulate had to be true. However, his definition of parallel and Euclid's are not the same, and that Euclid chose the one he did is one more tribute to Euclid's genius. First, Euclid's definition is better because it makes no mention of distance, and the fewer things a definition has in it, the better it is. Second, Posidonius's definition assumes more than Euclid's. If two lines are always the same distance apart, then they are parallel, since they obviously can never meet. But Euclid's definition says nothing about how far apart parallel lines are. Two lines are parallel as long as the distance between them is never zero; and as far as Euclid was concerned, the distance between them can increase, decrease, or wiggle up and down. If you make the hidden assumption (or unhidden assumption, as Posidonius did) that the distance between them cannot change, then the fifth postulate can be proved. But you have not reduced the total number of assumptions that you have made.

The reason that the fifth postulate is often called the parallel postulate is that it can be replaced by a statement about parallel lines. The fifth postulate follows from the assumption that through a point outside a line, one and only one line parallel to a given line can be drawn (see Figure 20). The reason is that if there is only one parallel, it will be the line that makes a right angle at P. Any other line, in particular those that make the sum of the two angles less than two right angles, will not be parallel to AB and hence will intersect it. This form of the fifth postulate, sometimes called "Playfair's Postulate," has even replaced Euclid's version in some books.

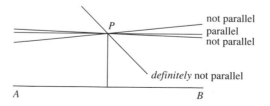

not parallel
parallel
not parallel

definitely not parallel

FIGURE 20 A B

There are many other statements that could be used as replacements for the fifth postulate. If any one of them could be proved, it would follow that the fifth postulate could also be proved. If you could show that

Two lines parallel to a third line are parallel to each other,

then you would have the fifth postulate proved. But this is like the common notion that two things equal to a third thing are equal to each other: if you want to have it, you must assume it. If you could establish that

The points equidistant from a line (on one side or the other) form a line,

then the fifth postulate could be proved. But you cannot prove that either. The fifth postulate can be derived from

Given any number, you can find a triangle whose area is larger than that number,

or from

Given a triangle, there is another triangle similar to it and not equal to it,

or from

There exists a rectangle.

It may seem strange that it is impossible to prove that rectangles exist when you can dash one off with a pencil at any time. But whenever you do and say "See? A rectangle!" you are unconsciously assuming the truth of Euclid's fifth postulate. What you *should* say is, "Assuming that if two lines fall on a straight line in such a way that the sum of the angles they make with one side of the straight line is less than two right angles then the two lines will intersect on that side of the straight line, here is a rectangle." However, if you say that, people will soon give you funny looks and avoid your company. Euclid's fifth postulate, and the things that imply it, seem so thumpingly obvious and intuitively clear that it is frustrating indeed that they cannot be proved.

In 1733, Gerolamo Saccheri claimed to have proved the fifth postulate in his book *Euclides ab Omni Naevo Vindicatus*. Saccheri constructed a quadrilateral whose top and bottom were parallel lines and that had right angles at the base (see Figure 21). If the angles at the top are right angles—that is, if you can draw a rectangle—then Euclid's fifth postulate follows. When mathematicians have no direct way to show that a statement is true, they are fond of supposing that the statement is false and then showing that the assumption leads to something that is obviously false. Since only truth can follow from truth, if falsehood follows from

FIGURE 21 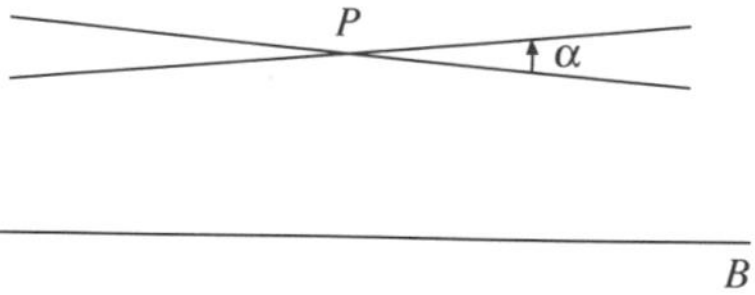   FIGURE 22

something, then the something from which it follows must be false. So, if you assume that one of the angles at the top of the quadrilateral is acute and find that it follows that $2 = 1$, then your assumption was incorrect.

That is the method that Saccheri used. He supposed that the angle at the top left is not a right angle. It then must be obtuse or acute. He showed that if the angle is obtuse, then something false follows. If the angle is acute, he showed that many other things would have to be true. Eventually, he showed that it would follow that there were infinitely many straight lines that would never meet a line (see Figure 22): *all* the lines that pass through P inside the angle α will never cross AB. Saccheri concluded that was "repugnant to the nature of the straight line," and hence that the top left-hand angle could not be acute either. That, however, is not the sort of conclusion that convinces many people. Saccheri may have realized the lack of force of his argument, since he later gave another proof that the acute angle is impossible, also unconvincing; and, in addition, his book was not published until after his death. Perhaps he did not have it published because he was waiting, hoping that he could find a better proof.

The history of the fifth postulate gives more evidence, as if any more were needed, that the collective human mind moves slowly. You might have thought that in the two thousand years since Euclid wrote the *Elements* someone would have said, "Hey, everybody, listen, I've got an idea! I bet that Euclid knew what he was doing when he made the fifth postulate a postulate—I bet it *can't* be proved! What do you think of that?"

If anyone said that before 1763, either no one was listening or no one took it seriously. It was in that year that Georg Klügel, a name insufficiently celebrated in the history of mathematics, finished his doctoral dissertation. Kästner, his supervisor, evidently decided that Georg lacked the talent to do anything original, and thus set him to work on a dissertation of the second class (mathematicians tend to hold mere scholarship in contempt): one critically examining the literature on the fifth postulate. Klügel examined twenty-eight attempts to prove the postulate, found them all wanting, and offered the opinion that it could not be proved and that people accepted it as true only because of the evidence of their senses.

Well of course, we now think, that was obvious. Obvious to us, but no idea can be obvious until someone has it for the first time. As far as I know, Georg Klügel was the first person to think it unnecessary that the fifth postulate be true, or that Euclidean geometry has to be the one true geometry, or that there necessarily exists a rectangle. All hail, I say, hail Klügel! After finishing his dissertation, he eventually became professor of mathematics at the University of Helmstädt and, though he no doubt led a useful life, he disappeared from the annals of mathematics. Kästner had judged his level of talent accurately.

Once an idea appears, people who might not have been able to conceive it in the first place are able to take it further, and this is what happened with Klügel's revolutionary thought. Johann Lambert (1728–1777), Ferdinand Schweikart (1780–1859), and Franz Taurinus (1794–1874) considered consequences of the falsity of the fifth postulate and concluded, hesitantly, that a geometry where there could be more than one line parallel to a given line through a given point might be logically conceivable. But, they were quick to add, of course Euclidean geometry was the only possible geometry of space.

The great Gauss as usual saw further and more clearly. He saw that a geometry in which there could be more than one parallel drawn to a line through a point was not only possible, it might even be the geometry of space. One of the theorems of that geometry is that the sum of the angles of a triangle is less than two right angles. So Gauss measured the angles of a huge triangle (its longest side was 67 miles long) and found that, within the limits of experimental error, their sum was 180°, so he could come to no conclusion about the geometry of space. Gauss never published his non-Euclidean geometry because he had plenty of other things to keep him occupied, and he was a man who liked a calm life. He thought that if he were to intimate that Euclid's geometry was not the only possible geometry, a huge fuss would be raised by people who put the *Elements* in the same class as divine revelation. He may have been mistaken, since when Nikolai Ivanovich Lobachevski (1793–1856) published his first work on non-Euclidean geometry in 1829, no one seems to have noticed. In his book *Geometrische Untersuchungen zur Theorie der Parallellinien*, published in 1840, he commented on the lack of interest in his research. János Bolyai (1802–1860), who worked independently of Lobachevski, published his work on non-Euclidean geometry in 1832 and no storm arose. In fact, it was not until after Gauss's death in 1855 and the publication of his material on non-Euclidean geometry that people began to pay attention to it—it needed to have the name of Gauss attached to it. But, at last, more than a century after Saccheri, the fact of non-Euclidean geometry was known.

Not only was non-Euclidean geometry known, it was shown to be every bit as good as Euclidean geometry. That is, it was shown to be as consistent as Euclidean geometry. A logical system is consistent when it has no built-in contradictions, so it is impossible to have both a theorem and its negative following from the postulates. For example, if you added a postulate 6 to Euclidean geometry, saying that in a right triangle with legs of length a and b and hypotenuse of length c, $a^3 + b^3 = c^3$, you would have created an inconsistent geometry, because you can prove from the first five postulates that $a^2 + b^2 = c^2$, which conflicts with postulate 6. Similarly, if you replaced postulate 5 with the non-Euclidean postulate that more than one parallel to a line can be drawn through a point, and added a postulate 6 stating that a rectangle exists, your geometry would be worthless since the first five postulates imply that no rectangle exists. In the nineteenth century several people, notably Henri Poincaré, constructed models of non-Euclidean geometry in which the postulates of non-Euclidean geometry are theorems in Euclidean geometry. Once you have such a model, it follows that if Euclidean geometry is consistent (and no one doubts that it is), then so is non-Euclidean geometry, because any contradiction in non-Euclidean geometry would translate in the model into a contradiction in Euclidean geometry. Thus we have Euclidean geometry and non-Euclidean geometry, completely equal as logical systems.

The question then arises, what *is* the geometry of space, Euclidean or non-Euclidean? The evidence of our senses is all in favor of Euclidean geometry, but our senses are not to be trusted: they tell us, among other erroneous things, that the earth is flat and the sun goes around it. The question is impossible to answer. If we want to, we can assume that space is Euclidean, or we can assume that it is non-Euclidean, and our convenience is the basis on which to make the choice. We cannot determine the nature of space any more than the inhabitants of a two-dimensional world could determine if the sheet they are living on is flat or curved. We could tell them because we are outside of their world and can look at it from the outside, but there is no way for us to escape our three-dimensional universe and look back on it. Nor have any four-dimensional creatures as yet seen fit to let us in on the secret. We will never know.

Most people are able to bear the uncertainty of not knowing what kind of space they are living in with equanimity, probably because not one person in a hundred thousand has ever considered the matter. But some cranks cannot. One characteristic of cranks is that they prefer the simple to the complex. The ratio of the circumference of a circle to its diameter should be something simple like 22/7 or 3.125 instead of 3.14159265358979.... Angles should be trisectable with straightedge and compass. There ought to be a simple rule for determining prime numbers. Thus, when cranks turn their attention to geometry, they assert that non-Euclidean geometry is a delusion and that St. Euclid had divinely inspired truth. It is a comfortable and dependable world, the Euclidean world, where there is only one line through a point parallel to another line, and where the angles of a triangle always sum to 180° exactly, so it is easy to see its appeal. Thus, cranks try to demolish non-Euclidean geometry, and the quickest way to do that is to prove the fifth postulate. A non-Euclidean world where straight lines do not look straight and there are no rectangles is too strange and confusing for cranks; they want their geometry to be clear, definite, and familiar. It is the same sentiment that makes grown-ups sometimes wish they were children again. However, adulthood is not reversible and we must all face the fact that we do not know and can never know whether or not space is Euclidean. All of us will have to die never knowing if we have ever really seen a rectangle. That may not be how we would like the world to be, but there is nothing we can do about it. If you try to do something about it, you are a crank.

The most amazing example of a non-Euclidean crank was J. C. (1878–1968), a member of the Roman Catholic priesthood and a college president from 1917 to 1940. In 1931, while president of Duquesne University (Pittsburgh, Pennsylvania), he published *Euclid or Einstein*, a 310-page hardbound book devoted to proving the fifth postulate and thereby demolishing Einstein's relativity. No other college president that I know of has ever proved Euclid's fifth postulate. No other college president that I know of has ever been any sort of mathematical crank. College presidents are chosen for reasons other than their mathematical knowledge, and they are usually much too busy doing other things to spend time on mathematics. Almost all college presidents are also much too busy to produce 310-page manuscripts, filled with footnotes and with quotations in *five* languages to introduce chapters (the quotations are from, respectively, Cicero, Goethe, Horace, Proclus, Horace again, Aristotle, Shakespeare, Pope, and Dante). The book has the appearance of being the product of a commercial publisher, though Devin-Adair, the company that brought it out, has disappeared. Cranks do not have books published by reputable publishers, cranks do not have classical learning at their fingertips, and cranks are usually not college presidents. C. was amazing.

His book displays great learning. He had read widely about non-Euclidean geometry, more widely than ninety-nine mathematicians out of a hundred, and he knew the history of the fifth postulate backwards and forwards. The first three chapters, on Euclidean geometry, the history of the parallel theory, and the history of non-Euclidean geometry, could serve as a textbook survey if the crankish bits were removed. However, crankery is there. For example, he says

> Mathematics is the science of measurement of physical quantity

(no, it isn't),

> Geometry is the science of possible determinations of space

(way off the mark), and

> None of the neo-Geometers or their followers ever thought of furnishing a proof for the contradictory of Euclid's postulate, although they are loud in demanding apodictic proof for it; they merely deny the one and

affirm the other. Such is the method adopted for the trumped up and loudly trumpeted non-Euclidean Mathematics of the moderns. Still to hear them, traditional mathematics is but a beginning as compared with theirs.

The excerpts show some of the working of C.'s mind: mathematics is about physical reality; geometry is about physical space; there can be only one physical space, and that one is Euclidean. Nevertheless, he assessed Saccheri correctly:

> Saccheri's is really the most unsatisfactory of all the attempts to establish the fifth postulate. It is a long drawn out *argumentum per impossibile*, by far the longest of all, which, in final reckoning, is no more satisfactory or convincing than any of the others. His attempt would have long ago faded into obscurity, were it not for the rise of non-Euclidean Geometry, and for the fact that non-Euclidean Geometers have come to regard him as one of their precursors. His work was in reality forgotten until it was again hailed forth by Beltrami in 1889 in a tract entitled, *Un precursore Italiano de Legendre e di Lobatschewsky*.

Footnotes in C.'s book refer to Engel and Stäckel's *Die Theorie der Parallellinen von Euclid bis auf Gauss*, to Heath's edition of Euclid's *Elements*, to *Euclid and his Modern Rivals* by the author of Alice in Wonderland, to Delaporte's *Essai Philosophique sur la Géometrie non-Euclidienne*, and to the *Irish Ecclesiastical Record*, among other sources. What a college president! We need more like him today. Like him, that is, in breadth and depth of learning and concern with matters of the mind. Crank college presidents we can do without.

It is natural to wonder why C. did not drop by his mathematics department or summon one of its members to inquire about the status of the fifth postulate and non-Euclidean geometry. In 1931 everyone knew that non-Euclidean geometry was every bit as good as Euclidean geometry and that the fifth postulate could not possibly be proved. We will never know why, and in fact it may be that C. did indeed consult with all kinds of authorities. However, I think that he did not and I think that I know the reason why. In 1931 C. had been a college president, not of colleges of the first rank, for fourteen years. That means that he had been dealing with college faculty members, not of the first rank, for fourteen years. Given C.'s high opinion of himself—no one without a high opinion of himself would attempt to overthrow non-Euclidean geometry *and* relativity—I think it is very probable that he had developed feelings of contempt for what he may have seen as the puffed-up but mediocre prima donnas on his faculty and could not bring himself to ask them for help. Even cranks who are not college presidents often do not feel the need for any information from outside.

Chapter 4 of his book contains the proof of Euclid's postulate. It is very clear, since it is the same as Posidonius's. C. said, straight out,

We shall also change Euclid's definition of parallel lines to correspond with our construction.

Parallel lines are lines that lying in the same plane are equidistant at equidistant points.

From this, it follows quickly that there are parallelograms and rectangles, that the sum of the angles of a triangle is two right angles, and the fifth postulate is proved. All done! Except that definitions may not be changed at will, as C. realized:

We shall omit all discussion of this definition till we treat the question of parallel lines by itself, when we propose to justify it completely.

Chapters 5 and 6, "The Parallel Theory" and "The Definition of Parallel Lines" are where C. tried to show that his definition of parallel lines is exactly equivalent to Euclid's, saying neither more nor less. On the face of it, C.'s definition says more, since if two lines are everywhere equidistant they can never meet; and we know that it in fact says more since with it the fifth postulate can be proved, while with Euclid's definition it cannot. He started, on page 94, with:

There are two questions concerning the parallel theory that should occupy our attention. One is the question of the proper definition of parallel lines, which, as we have seen, has been a matter of dispute since Euclid's time.

He concluded, forty-five pages later, with:

We have now completed the demonstration of the definition of parallel lines and the postulate of Euclid. . . . We did this without using anything except what could be deduced from the other postulates and theorems universally accepted. Euclidean geometry is therefore completely established.

It is in these forty-five pages that someone searching for error must look to find where C. went wrong in his demonstration that the two definitions of parallel lines are actually the same. The searcher will have a tough time. C. was a tribute to the Jesuitic system of education of which he was a product. He reasoned closely, and those without similar training are at a disadvantage. The fifth postulate says that if you have two lines making with a third a sum of angles less than two right angles, then the lines will meet. That is, they will form a triangle. The converse of this is that if you have a triangle, then the sum of any two of its angles must be less than two right angles. This is a theorem that Euclid proved in the *Elements*. "What of it?" you may think. Everyone knows that the truth of a theorem does not imply the truth of its converse—from the truth of "If I were rich, then I would not work" the truth of "If I do not work, then I am rich" does not follow. Ah, but see what C. made of it:

Euclid himself proves the converse of this postulate in I, 17, where it is shown that in any triangle two angles taken together in any manner are less than two right angles. If the converse requires proof, then the postulate itself is a theorem that requires demonstration. For a postulate should be such that the connection between subject and predicate ought to be clear and self-evident no matter what way they are turned. If there is an essential and evident connection between the formation of triangles and base angles that together are less than two right angles, then both the proposition concerning this connection and its converse will be evident. The fact that Euclid proved the one, and assumed the other, can only point to one thing, his inability to furnish proof for the other. But in the proof of I, 17 Euclid does not give the actual cause of the connection; it is rather a proof that the connection exists. For the reason does not lie in the relation to the exterior angles, but in the necessary convergence to a point that is in the sides of a triangle in the case of two lines forming with a base less than two right angles. Even I, 17 is then not a proper demonstration. It is what Aristotle calls demonstration ὅτι and not a demonstration διότι which is a proper mathematical demonstration. For there is a great difference between knowledge "that" a thing is, and "why" it is. That Euclid had recourse to this shows that he was hard put to prove his theory. But it shows at the same time, that the theory required proof, and that for the postulate as well as for its converse.

I could not follow all of that, but what C. seemed to be saying is that if the converse of a postulate is a theorem, then the postulate can be proved. To blast that reasoning out of the water you would have to find an example of a mathematical system, other than Euclidean geometry, that contains a postulate whose converse is a theorem, but in which the postulate could not be proved. Can you think of one offhand?

And so the forty-five pages go. There is no place where someone searching for error can say, "Aha! *Here* is the place! This step, from line fifteen to line sixteen, is not right, as the following counterexample shows." There is only the descent into the fog,

There are two questions concerning the parallel theory that should occupy our attention,

and the triumphal ascent out of it, pages later, waving the prize overhead:

Euclidean geometry is therefore completely established.

In between is such stuff as:

But we must distinguish between faulty and incorrect definition; between the actual definition or explanation of a thing as it compares with the rules of logic, and the actual content of the definition as representing reality and truth. It may be that a certain definition is faulty in logic, but the conception back of it which the author attempts to put into words, may be correct. So long as the latter is true, and the truth is kept in sight, there is no danger to the science. The defect in the definition can be easily corrected. As an actual fact, Euclid's definition has a correct idea back of it, and the actual wording need not worry us very much. The correct form can easily be given. He accepted the form doubtless to tally with the assumption in the fifth postulate. But the definition expresses a real scientific property of parallel lines, and it could very easily be drawn as a conclusion from any real definition.

Fog cannot be pinned down and neither could C. Fog is the reason so many cranks can maintain, with some degree of truth, "No one has ever found anything wrong with my work." Chapters 5 and 6 are forty-five pages of fog.

After proving the fifth postulate to his satisfaction, C. had the fun of blasting away at non-Euclidean geometers and relativists. What fun to pick the apposite quote from *A Midsummer's Night's Dream* to head the chapter called "The Verdict on Metageometry"!

A play there is, my lord, some ten words long,
Which is as brief as I have known a play,
But by ten words, my lord, it is too long,
Which makes it tedious; for in all the play
There is not one word apt, one player fitted.

What fun to sneer at Einstein!

Einstein has such a faculty for embracing both sides of a contradiction that one would have to be of the same frame of mind to follow his thought, it is so peculiarly his own. His thought is but odds and ends, unconnected bits, incongruous, undigested, and contradictory. Whenever he is as a pure mathematician dealing in pure mathematical symbols, he becomes the most out-and-out careless thinker the moment he gets beyond his symbols and his equations. He is guilty of the very vice he attributes to Euclid, that of spinning a web of the pure logical-formal, with no relation to reality; and then when he tries to establish contact between this and reality, his thought staggers, and reels, and stumbles, and falls, like a blind man rushing into unknown territory. . . .

Even worse than this is the absence of the power of criticism to enable him to see the glaring contradictories which he is embracing, and the lack of logical insight to understand the use and force of language. No wonder then he blunders, and in floundering, grasps for odds and ends left him by his predecessors in the same wild school of thought.

What fun to condescend to Riemann!

Riemann was also a young man of twenty-eight when he read the now well-known *Habilitationsschrift* to the Philosophical Faculty of Göttingen. It is rather a jejune effort written in ponderous language, and is not remarkable for any striking originality either in questions of geometry or on the metaphysics of space. . . .

It is not merely a question of not being practiced, but of complete incompetence. He is not entitled to indulgent criticism, for he had no business meddling with what he knew nothing about. A prudent, sensible man would not get beyond his depth and would stick to what he was competent to handle. His lack of knowledge and ability to cope with the matter he undertook to illustrate in such a way that he was going to do away with what he terms the darkness of two thousand years, has been the cause of introducing the grossest errors into science. Mathematicians, who are not philosophers as well, should learn to stick to their lasts.

The last paragraph is remarkable. Interchange "philosophers" and "mathematicians" in its last sentence and the paragraph applies not to Riemann but to its author, the Very Reverend J. C., S. J., President of Duquesne University, the self-proclaimed superior of Einstein, Riemann, Lobachevsky, Bolyai, Gauss, and Euclid.

The cause of C.'s dislike of non-Euclidean geometry was, I think, his belief that mathematics and geometry have to be about physical space, and that they have to be *true*. That theorems have the form "If A is true, then B is true" and not "A is true, therefore B is true" did not make an impression on him. He attacked Riemann for asserting that n-dimensional manifolds can exist—they are not *true*:

The only triply extended manifoldness, or to use the commoner combination of terms, three-dimensional magnitudes, is that extension of bodies represented by our idea of space. Yet Riemann asserts (again it is pure assertion for which there is neither proof nor illustration) that many other three-dimensional magnitudes are conceivable. He attributes their creation and development to the higher mathematics. We have already seen how he constructed them. They are not matters of fact or experience but a pure arbitrary creation without even a definable concept back of them.

He defended Euclidean geometry—because it is *true*:

There is not even the wraith of a positive reason for rejecting the postulate of Euclid, for no reason capable of creating positive doubt has ever been brought against it. The arguments that have been adduced are sophistic, and are invented to bolster up a negative doubt that arose purely from inability to prove a proposition that was very nearly self-evident; but this want of proof was more than offset by the complete lack of competency to show even any manner of plausibility for the contrary proposition. . . .

All the authority of common sense, all the weight of the accumulated and personal experience of the human race, the pragmatical sanction of a wonderfully coordinated and thoroughly consistent body of science that has satisfied all the practical and theoretical needs of the human spirit in those things for which the science may be used (Art and Science are full of it. Parallels, squares, right angles, are used by engineers, architects, surveyors, etc.), all are tossed aside for a mere nothing, for a contradiction out of a spirit of petulance at not being able to prove what all men have always held to be true. Such evidently is not the sign spiritual of intellectual benefactors of the human race.

Non-Euclidean geometry leads to relativity, relativity leads to a four-dimensional model of the universe, and that leads to something unacceptable:

If the universe, as Einstein holds, is a four-dimensional continuum, then there is no such thing as succession of time in a three-dimensional continuum. Every event in the world is synchronous just as all parts of a continuum are co-existing at the same time. The four-dimensional continuum is either actual or it is not. If it is not, there is no such thing as an actual four-dimensional universe. If it is actual, then there is no such thing as succession in the universe; it is altogether at once, just as two portions of space co-exist together, as New York, London, Paris, Rome. This is recognized by Eddington, the well-known English exponent of Einstein's Relativity. He says: "Events do not happen, they are just there, and we come across them."

In such a case everything would have to be really simultaneous, and change is only a situation in a different spot of the space-time continuum. . . . It would be impossible for a falling body to change its position in a three-dimensional continuum in successive moments of time, since there is no such thing. The whole change would be a mere static condition of the wonderful space-time. The notion of succession would have to be dropped out of time, and therefore all actual change, even of successive positions on space-time, is impossible.

C. did not realize that what Einstein and physicists do is construct models of reality that may be helpful in explaining observed phenomena. A model of a thing is not the thing. The map is not the territory.

There are some other implications contained in this wonderful theory. For instance, we never change, we are not really born, nor do we die; we are simply spread out over a portion of space-time like some huge sprawling monster. This is recognized by Eddington, who at least is not afraid to deduce the conclusions from Einstein's Relativity. "An individual is a four-dimensional object of greatly elongated form; in ordinary language we may say that he has considerable extension in time, and insignificant extension in space. Practically he is represented by a line—his track through the world." Just try to get a mental photograph of yourself to see what you look like when you have a figure of the three ordinary dimensions and one other dimension of three hundred thousand kilometers for every second of your existence.

Eternity, too, is reduced to a very simple thing. We are at least as changeless and eternal as the four-dimensional world of which we form a portion. The poet's conception of time that never returns becomes false, for it never goes away. No part of time flows, any more than New York flows, or London flows, or Peking flows. It simply remains where it is. The old metaphysical religious questions that have long troubled the world are all solved. We are immortal: eternity is with us, for we never change. Since all the events of our life are but simple positions in space-time, they are there as they always were. Our consciousness is just playing us a prank, when it tells us of change and growing old; and our intellects deceive us when they represent life and death.

Could this be the real reason for C.'s proof of Euclid's Fifth Postulate? His proof would make non-Euclidean geometry go away (he thought), and so (he thought) would relativity, that system that implies (he thought) that we live in an unchanging universe in which there can be no free will, hence no need for religion, hence no need for priests, and hence C. has lived his life in vain. Imputing self-interest as a motive to someone whose ideas you do not agree with is not playing the debate game fairly, but I think that C.'s last two paragraphs explain a great deal.

However, they do not explain a later work. *The Foucault Pendulum and the Newtonian Theory* is a 34-page pamphlet published by one of C.'s nephews in 1975, after the author's death in 1969, in his ninety-first year. It is not clear exactly when it was written, but it was sometime after 1963. C. wrote:

I think it was in the spring of 1902 that I witnessed a reproduction of the Foucault pendulum experiment, in the place where it was first made, under the cupola of the Pantheon of Paris. A professor of the Sorbonne explained it daily to the few that were interested in it. The explanation failed to satisfy me, and I found that when I brought forward an objection, it was not exactly received with affability. A few years later the same experiment was repeated at Columbia University, and was quite copiously explained in the old New York *Sun*. My recollection is that it was the same explanation as that given in Paris. I thereupon set down what I thought wrong with it, along with what I held to be a true explanation. Its fate was the scrap basket. I had my first few lessons about the stodginess of scientists, who do not like to be told they are wrong, especially by a young unknown. Who are you anyway? I had read what Emerson said about mousetraps and must have believed him.

After this experience, the idea lay more or less dormant until the Russians sent up their first sputnik. I knew that neither they nor ourselves could pinpoint space trajectories because neither had a correct physical basis for a gravitational theory. I wrote to the then Senator Johnson, now President, who was chairman of the Senate Committee in charge, explained my view, and made two requests: one was that, for scientific purposes, the laws of induction should be applied to our space experiments; the other was that the Foucault experiment be renewed, to learn more about the gravitational field. The Senator became interested and took the matter up with those concerned, but he was told they could do nothing without the O. K. of the American Foundation of Science, which was evidently an easy way of disposing of the question. Though I was aware of the futility of doing so, I wrote to the Academy making the second request, and received a polite letter, telling me it was placed on file, etc. So I was disposed of.

Not quite. By that time I was getting a little weary of receiving such replies, equivalent to telling me: "Run home, little fellow, and play in your own back yard; you don't belong in this league." I agree, I don't, but it is the other way around; they don't belong in mine, not by a thousand miles.

The pamphlet contains physical rather than mathematical crankery, but it is an interesting footnote to C.'s *Euclid or Einstein*. First, it shows that his experience with that book did not extinguish his desire to set the scientific world straight on matters on which he thought everyone except himself thought incorrectly. Second, I thought that C., after exterminating Einstein, would have wanted to return to the comforting simplicities of Newtonian physics, but his pamphlet asserts that Newton, too, was mistaken. C. found ten errors in physicists' treatments of the Foucault pendulum, not one of which I could understand. Here, for example, is

Error No. 4   This is an error of fact. We are dealing with the physical earth and not with Newton's mathematical notions. Two forces are actually involved, rotation and gravity. They are both there from the beginning. Gravity arranges the line of the pendulum such that the latter shall be in equilibrium when this line points to the center of gravity. Since every part rotates with the earth, the rotation will keep the pendulum in continuous equilibrium as long as the center of gravity coincides with the center of rotation, so that the line of the pendulum is the continuation of the radius. In this case, the line of parallel planes is completely impossible, for these are the very forces that make the plane of the meridian. If this is the theory held, and it is that of Newtonian physics, scientists must give up their explanation of the Foucault pendulum, which it flatly contradicts. Of course, they would never be so logical as that; they want at least the semblance of having their cake and eating it.

The second part of the pamphlet is an attack on the theory of the tides, in which C. found twelve "Ineptitudes." He said:

There is not a shred of evidence that the moon pulls anything, not even a spoonful of water one millimeter in height.

It was not clear what C. wanted in place of Newton's ideas. Perhaps it was those of Aristotle.

Euclid's "World-Renowned Parallel Postulate" by M. R., a 30-page booklet published in 1905, involves religion more directly with the fifth postulate than did C.'s book. R. summarized part of its contents:

Section I.   Exposes the arrogant and satanic pretensions of the non-Euclideans. And shows the infinite authority of Theorems A, B, C.

Sec. II.   Extracts from the false report on non-Euclidean geometry by "The Am. Assoc. for Adv't. of Science."

Sec. III.   Reveals the dark understandings and bewildered minds of the non-Euclideans.

Sec. IV.   Fundamental Theorem, demonstrating the homogeneous simplicity, infinity, immutability, and divine nature of Euclidean Space. Also showing the *contradictory* teachings of the Satanic "non-Euclideans."

Sec. V.   The study of "non-Euclidean" Geometry brings nothing to students but fatigue, vanity, arrogance, and imbecility.

R. was convinced that God had ordained that space be Euclidean, that non-Euclidean geometry was the invention of Satan, and that non-Euclidean geometry implied the truth of the false doctrine of evolution:

The Satanic anti-Euclideans hope to discover a triangle whose 3 angles will differ from the sum of two right angles; for then they would claim the "*non-Euclidean*" nature of space; also, the proof of Darwinism.

R.'s proof of Euclid's parallel postulate is non-Euclidean, since it has lines moving around, not something that Euclid allowed. His proof contains reasons like:

Reason 1.   All motion, whatsoever, is *continuous* and *successive*; and therefore, CA, in its rotation cannot pass over any angular space, as CAG, GCH, etc., without previously passing over in regular *successive order*, each of the *infinitesimal* parts of such spaces.

Reason 2.   Because the continuous left alternate *intermediate* series of angles described in the foregoing *Enunciation*, occupy or fill up completely the whole angular space ACF; therefore (per Reason 1), among this continuous series of intermediate angles no vacancy is left for the existence of any other intermediate angle, whatsoever; and wherefore, all the just described series of continuous decreasing angles can have no other *common difference* than an equal *infinitesimal* one.

Euclid would not have approved.

R. found a mathematician who tried to set him straight:

Extracts from letter of [the mathematician], dated Feb. 3, 1902.

(a) "To prove Euclid's postulate by his other hypotheses is as *contradictory* as to prove that 2×2 equals 5."

(b) That, "if we discard the question of the physical space in which we live, an infinite number of geometrical systems can be invented."

(c) "That the propositions deduced from these geometric systems may be *contrary* or *contradictory* to each other; and yet, it would be a misunderstanding to think that either of these systems was wrong."

Here is how R. demolished him:

In (b), foregoing, the reference to "physical space" is *absurd*. Space is neither a *material* nor *immaterial* substance. And, because non-Euclidean geometry is founded on a *false assumption*, therefore, "an infinite number of (*false*—[M. R.]) geometrical systems can be invented," as easily as the endless number of false religions, etc.

In (c), foregoing, the eternal "*Principle of Contradiction*"—the foundation of all knowledge—is denied.

Thus, "Everything either is, or is not." "A thing cannot *be* and *not be*, at the same time."

An Infinite *Axiom*: Whatsoever doctrines are founded on *false principles* are necessarily *false*.

R.'s concluding words were:

1. The crazy Satanic deductions from Non-Euclidean geometry; together with the foolish declarations of the Non-Euclideans, show the invention of the "imaginary" or Non-Euclidean Geometry to be beastly foolish; and, therefore, of a nature to bewilder the minds, to squander the time, and to destroy the health of millions of students every year.

2. The teaching of "imaginary" or Non-Euclidean Geometry in colleges and schools would breed an arrogant and imbecile race of students, who would endanger society by the application of "imaginary" and fallacious reasonings to the most important subjects, such as human government, labor and capital, Christian doctrine, Christ's miracles, God.

Let Euclidean schoolboys and professors, and also all sincere philosophers (both men and women) throughout the world, rejoice, because the Eternal Euclidean Geometer—"who commands the stars in their (Euclidean) courses"—hath revealed to the humble author hereof the two irrefragable and *a priori* demonstrations of the Euclidean "world-renowned parallel postulate," as the reward of 47 years of contemplation.

FOUR-COLOR THEOREM, THE

If you draw regions in the plane and label them with numbers so no two adjacent regions have the same number, you will find that you need only four numbers (see Figure 23). You cannot do it with fewer than four (see Figure 24). You might think that by taking the first diagram and altering it so there is a new region that touches old ones numbered 1, 2, 3, and 4 that you would get a diagram that needed five numbers, but it is always possible to renumber it so that four numbers will do, as in Figure 25.

To seize the interest better, the problem is usually put in the form of asking how many colors are necessary to color a map so that no two countries having a common boundary have the same color. As the problem in this form was passed on, a folk legend grew up that mapmakers had made the discovery that four colors were enough. The fact that inspection of any decent map would show that invariably five, six, or even more colors were used did nothing to discourage its spread. Even the

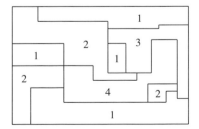

FIGURE 23

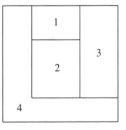

FIGURE 24

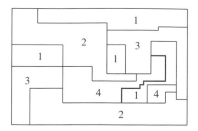

FIGURE 25

fact that K. O. May and H. S. M. Coxeter determined that the origin of the problem had nothing to do with maps or their coloring has not kept it from dying out. The legend must fill some deep need.

The actual origin of the four-color theorem (or, the four-color conjecture, as it should be called before it was proved) was an observation made in 1852 by Francis Guthrie, a student at University College, London, that four colors would be enough to color a map of the counties of England. The same observation may have been made before by other people, possibly even by some mapmakers, but if so nothing came of it. Francis Guthrie mentioned his observation to his brother Frederick, and Frederick mentioned it to his teacher, Augustus De Morgan. De Morgan, a great letter-writer, then mentioned it to W. R. Hamilton, the discoverer of quaternions:

> A student of mine asked me today to give him a reason for a fact that I did not know was a fact, and do not yet. He says, that if a figure be anyhow divided, and the compartments differently colored, so that figures with any portion of common boundary *lines* are differently coloured—four colours may be wanted, but no more. Query: cannot a necessity for five or more be invented? . . . Now, it does seem that drawing three compartments with common boundary, two and two, you cannot make a fourth take boundary from all, except by enclosing one. But it is tricky work, and I am not sure of all convolutions. What do you say? And has it, if true, been noticed? . . . The more I think of it, the more evident it seems.

Yes, it does seem evident, and clearly no heavy mathematical machinery like partial differential equations is going to be needed to prove it. It is tantalizing, but it is also tricky. Hamilton declined the challenge:

> I am not likely to attempt your "quaternion of colours" very soon.

The conjecture, having been made, did not disappear, though nothing was heard about it until 1878, when Cayley told the London Mathematical Society that he had been unable to prove it. Mention by the great Cayley was enough to start people working on it, and in 1880 A. B. Kempe published a proof. The proof illustrates that the papers of mathematicians, sometimes reading as if they had been composed by a minor deity, are the productions of fallible humans, for it was wrong. The error was not noticed by the paper's referee or referees, nor by any of its readers; and as far as the mathematical community was concerned, the four-color conjecture was now the four-color theorem and settled once and for all. It was not until 1890 that P. J. Heawood read the proof critically, found that it was flawed, and that the flaw could not be removed. The theorem became a conjecture once again.

Evidently, it is not easy to tell when the conjecture has been proved. As H. S. M. Coxeter said,

> Almost every mathematician must have experienced one glorious night when he thought he had discovered a proof, only to find in the morning that he had fallen into a similar trap.

The four-color conjecture was easy to state and easy to understand, no large amount of technical mathematics is needed to attack it, and errors in proposed proofs are hard to see, even for professionals: what an ideal combination to attract cranks! All it lacked (thank heaven) was a prize offered for its solution. Thank heaven also that Kenneth Appel and Wolfgang Haken proved in 1976 that the conjecture is true. Now that the four-color theorem is a theorem again it is no longer a challenge to cranks. Because of the fuss about Appel and Haken's proof depending on a computer to check a mass of cases, crank interest in the theorem has not entirely died out; but in a generation or two the fuss will be forgotten and cranks can move on to something else. The proof is done.

However, when a theorem is proved, there is always the possibility of improving on it. This is what A. P. tried to do in "Plane maps are 2-colorable." The title is startling, but his idea is easily stated. If you want to color a map, you do not need four colors. You do not even need three: two will do. For example, take yellow and blue. Color some countries yellow, some blue, mix the two colors to color some green, and leave some white and uncolored. That's all, though P.'s manuscript runs to ten pages.

You would not think that such an observation could generate a thick file of correspondence, but it did. P. was deep into the literature of the problem, mentioning Wilf's theorem, Saaty's book, Ore's construction, Euler's relation, Franklin's derivation, Tait's equivalence, and, of course, Appel and Haken's proof. He was undeterred by responses such as the one that started:

> No, don't send me your paper. When mathematicians say four colors are necessary for coloring all planar maps, they mean that they are necessary for *some* maps.

Nor was he deterred by the one that included:

Thus, it is a word game or a game of interpretation of how to go from four to two that is being played here, not any mathematical theorem that is being proved. In fact, taking your approach to its logical conclusion, the plane is 1-colorable, since with sufficient thinning, I can get four distinct shades of one color to paint any map.

The first paragraph of P.'s response to the last letter is:

It is sometimes difficult to distinguish between a quibble and a subtlety. A number may have various forms; and I think it can be a subtle matter to determine which form is appropriate in which context. In deciding which form of 24 is appropriate in the map-coloring context, the cubic map of 3 area has to be taken into consideration along with the knowledge that there are at least 24 ways of coloring it.

P. carried on, at length, in the same vein. That was not his last communication, either.

Proving the four-color theorem has turned more professionals into cranks than other problems. R. C., the author of *The Four Color Problem*, an 8-page pamphlet and "A Publication of the [C.] Research Institute," was the holder of a PhD degree in mathematical physics, the author of an aerodynamics text, and the head of the physics department at a small college. He published his proof of the four-color theorem in 1977, just too late to beat Appel and Haken. His argument was clearly presented, but he dealt with the building up of maps whereas a proof of the theorem should start with a given map. That he belongs in the class of cranks is evidenced by another of his publications, *The Eternity Equation*, that was advertised on the back cover of his pamphlet:

This new technology can be found today! The theory that can lead the United States out of its state of dependency [on foreign oil] can be found in a new book, "The Eternity Equation" by Dr. [R. C.]. A patent on a Resonance Absorber, by which the release of nuclear energy can be achieved, has already been granted under the title, Low Temperature Heat Transfer Device, and assigned U.S. Patent Number [patent number] and fully described in the book.

Like his proof of the four-color theorem, his resonance absorber seems not to have caught on.

C., though a professor, was not a professor of mathematics, nor was his college one that anyone would call distinguished. In contrast, Professor J. T. was indeed a professor of mathematics, at a college of high prestige, and he had achieved some eminence in his profession through writing and editing. However, such is the difficulty of the four-color theorem: after he had proved it to his satisfaction he could not see why no one else would agree that he had proved it, and so he spent his declining years unhappily in crankdom.

T., who was born in 1898, published his "The four-color theorem," a 13-page proof, in 1969. It was followed by a revised edition in 1971, and "A brief proof" in 1976, just before Appel and Haken produced a proof that was accepted (though perhaps not by T.) as being a proof. About his original proof, T. said:

> The work was privately printed. I offered it to no mathematical journal, not even to that which I founded and edited for eleven years, because the whole process of refereeing and editing, irrational at best, would be impossible in this case.

> I was entirely conscious that distribution would present difficulties. On October 1 last copies were mailed to a selected list which included the names of all with a paper reviewed by "Mathematical Reviews" in the last ten or twenty years. I hoped that the recipients would mention the availability of free copies, but the response was negligible.

He then took the step of advertising in the *Notices* of the American Mathematical Society. Advertisers in the *Notices* are usually publishers trying to sell their books and periodicals, or departments of mathematics trying to fill their vacancies. However, the Society will accept money from others wanting to use their pages. The quarter-page advertisement was not what a professional writer of advertisements would have produced:

THE FOUR COLOR THEOREM

> In 1898 P. J. Heawood gave a restricted system of linear congruences whose consistency implies the four-color theorem. That system is immediately equivalent to what might be described as the simplest non-trivial quadratic system of equations X. A solution of any given X can be easily had either directly or be specialization of general methods in the literature at that time. A pamphlet just published proves X consistent by applying mathematical induction to such solutions of subsystems S.... A copy may be had gratis from
>
> [J. T.]

It was not an advertisement that would stir many people to action. The *Notices* at the time had a circulation of around 15,000, and T. later wrote:

> [The advertisement] brought about 80 requests, but the flurry seems over. There is something here I don't understand. Incidentally, each of the 80 cost about 2 dollars just for the giving away process!

One of the things that he did not understand is that there is no reason for anyone to want a copy of something that, with probability .999 . . . , would repay any effort spent on it with nothing at all, except for the pale satisfaction of recognizing error. Teachers of mathematics get quite enough of that already, grading their students' examinations.

T.'s proof was reviewed in *Mathematical Reviews*, naturally with the reviewer pointing out an error. After corresponding with the reviewer for a time with neither side getting anywhere, T. was reduced to:

> I could find out nothing about him, the index of [*Mathematical Reviews*] listed no paper under his name, he seemed incapable of constructing a single "flawless" paragraph....
>
> I concluded he was crazy.

T. then complained to the editor of *Mathematical Reviews*, who replied:

> I'm sorry that the review did not meet with your approval. Our job is to produce more than 18,000 reviews a year and we cannot guarantee that each one is perfect. However, as far as I can determine, we did the best we could on the best advice that we could get.

The editor left himself open for T.'s comeback:

> To rephrase matters, how would you judge a store manager who said "Our job is to serve more than 18,000 customers and we cannot guarantee that no one is cheated"; or a newspaper editor who said "Our job is to serve more than 18,000 readers and we cannot guarantee that no one is libeled."

Nettled, the editor answered:

> There can be no discussion on the basis of your letters. Incidentally, my remarks about 18,000 reviews a year were meant to indicate that the editors cannot vouch personally for the accuracy of every review.... Therefore errors will sometimes occur but, given good will and cooperation on the part of authors and reviewers they can be caught and corrected. It saddens me to see such words as "cheated" and "libeled" in what purports to be a scientific discussion and I am afraid that the pressures on my time do not allow me to waste it on replies to any more communications of the same kind.

T., of course, had the last word:

> Anything I write is aimed to the point and I am prepared to substantiate it. Actions speak louder than words. Now that you have revealed your true attitude, your breaking off the discussion suits me very well. You have had your chance. This does not mean that I intend to let the matter drop.

He added, in the document in which he published the letters:

> I can readily understand the authenticity of the foregoing copies being doubted: how can one so stupid and dishonest be an officer of the Society? I shall be glad to furnish notarized copies on request. (A friendly tip to potential burglars: the originals are on safe deposit.)

As that parenthetical remark shows, T. was following the crank's path to obsession and paranoia. In 1974, he published *A Primer on Roots*, whose Chapter XIV, "Historical and Critical Remarks," could not leave *Mathematical Reviews* alone:

> The motivation behind the founding of MR was revenge in racial warfare, not mathematics. The fanatics responsible, backed by Oswald Veblen, would not listen to reason. It is sheer madness to waste effort and resources by having even two such journals, especially if one slavishly copies another thus renouncing the only conceivable reason for its existence, independence. It has neither the money nor, what is worse, the man-power to carry on its activities properly: by his own admission the job is too much for [the editor of *Mathematical Reviews*].

He also gave details of a separate controversy, that resulted in

> a circular letter sent by me as an individual to the roughly 27,000 entries in the combined directory of the three mathematical unions based in the United States. . . . Of the 77 replies received, 65 were not openly hostile. . . .

Exasperated, an official of the Mathematical Association of America was led to write:

> Fortunately, many members of the MAA are aware of the fact that for several years Professor [T.] has spent all his time attacking many professional organizations and institutions (including his own) with all kinds of unfounded charges and allegations.

T.'s riposte was:

> In a supposedly mathematical discussion this diatribe was written by a supposedly mature, educated [official] of an organization supposedly devoted to mathematics. Isn't mathematics sick unto death?

Naturally, there could be no improvement. At a meeting in 1975, the mathematician who had reviewed T.'s paper for *Mathematical Reviews* gave a talk. He said of it,

> There was an emotional scene at the end due to the presence of [T.], the one-time editor of [a mathematical journal], and author of several

books of expert mathematics on algebraic systems and related systems. The man has taken a personal crusade defending his proof of the 4CC (which I reviewed negatively in the *Math. Reviews*, citing the error) and turned it into a vendetta. He has now issued his (several times revised) proof in a new privately published book.... So far, no one except [T.] claims I am wrong, and he refuses to discuss the matter as a mathematical difficulty, in spite of lip service to mathematics, preferring to regard me as "crazy!"

T.'s version of the scene included:

The suggestion of "fisticuffs" is absurd: in 55 years I have never before heard the like suggested in connection with the meetings of the American Mathematical Society, which, even if irrational, are at least dignified.

Nothing would ever quiet T. or give him any peace except death, which arrived for him in 1978.

GÖDEL'S THEOREM

When Morris Kline titled one of his books *Mathematics: The Loss of Certainty*, he certainly hit the nail on the head. Things have been going downhill, certaintywise, for two hundred years. The golden age, when everyone was living in Eden, was after Newton had explained the universe with his laws of motion; when Kant could maintain that mathematics, in particular Euclidean geometry, could provide truth about reality, all by itself; when Laplace could answer Napoleon's question about the lack of any reference to God in his *Mechanique Céleste* with "I have no need of that hypothesis." The good old days! But then mathematicians, ever seeking truth even if it would do them harm, found that non-Euclidean geometries were every bit as good as Euclidean geometry, so that geometrical truth no longer implied physical truth. One bit of certainty gone.

Even if mathematics could not give us certainty about the physical world, there remained the hope of showing that mathematics itself was certain. Around a hundred years ago, Euclidean geometry was axiomatized properly, Whitehead and Russell were deriving all of mathematics from logic, and the haziness of set theory, another foundation on which mathematics could be based, was being cleared away. Hilbert wanted to make mathematical physics an axiomatic system. If we could only get everything organized as deductive systems, and if we could then prove that our systems were consistent and complete, then we would have the certainty of knowing that mathematical truth was under control. The physical world might dissolve into chaotic fog when its underpinnings were examined closely, but our mathematical world would be solid.

Alas, that hope was dashed also when Kurt Gödel showed that no mathematical system can demonstrate its own consistency, nor will we ever be able to know all of mathematical truth. So all mathematicians who are trying to prove theorems are not standing on a rock of certainty: they are standing on a large balloon that, for all they know, may pop any day now. Most of them are confident of the structural integrity of the balloon (and many do not even notice that it is not as solid as a rock), but a balloon it remains.

Removers of certainty are favorite targets of cranks, who prefer to live in as certain and understandable a world as possible. Thus cranks would like to show that π is a rational number, non-Euclidean geometry is a figment of perverse minds, and there must have been something wrong with Gödel too. Gödel's theorem does not get as much attention as π and the rest, because what he did is not generally known or understood; but he has not gone unnoticed.

The Foundations of Mathematics, published by G. M. in England in 1974, is a 12-page printed booklet headed

REVIEW MONOGRAPH (MATHEMATICS) No. M1

and subtitled

A New Analysis Showing Gödel's Theorem Based on Fallacy.

M. said of his monographs:

This is a series of occasional publications making known the results of fundamental researches. The main reasons for the existence of this independent series of publications are the lengthy delays and uncertainties following the submission of such articles to the established scientific and mathematical journals. Also, it seems desirable to publish these articles in a form allowing articles on a particular topic to be readily collected together and stored in a loose leaf binder, so that anyone wishing to read the research literature on that topic will have the literature readily to hand.

It is more likely that the real reason for the existence of the series was that no journal would print M.'s work. It is also likely that the first in the series was also the last, since the last sentence of the booklet is:

But, to continue this work requires money, and the author of this initial work appeals, therefore, for a funding of further work of such potential benefit to mankind.

Here is how M. explained the flaw in Gödel's proof:

In Gödel's 1931 Theorem, each ONE of a set of mutually exclusive logical elements is co-ordinated on a ONE-TO-ONE basis to each ONE of a set of arbitrarily selected numbers 1, 3, 5, . . . But, the Theorem is fundamentally fallacious, because in accordance with the theory on which it is based, these numbers co-ordinated to mutually exclusive elements are *not* distinguished in meaning from numbers meaning arithmetical inclusion, such as

$$2 = 1 + 1, \text{ as operative in } 2^3.$$

The fallacy in the Theorem may be shown by considering that the logical element "or" is coordinated in the Theorem to the number 7, which is not distinguished from $7 = 5 + 1 + 1 = \ldots$ But, 5 is co-ordinated to "not", 3 is co-ordinated to "the successor of" and 1 is co-ordinated to "the number 0". The logical elements "or", "not", "the successor of", etc. are each mutually EXCLUSIVE, whereas the numbers 7, 5, 3, 1 are as arithmetical numbers PARTIALLY INCLUSIVE of each other, the inclusions denoted by 7, 5, 3, 1 partially overlapping each other.

The best that can be said of this is that it is not to the point, and the worst is that it makes no sense at all. There is something about the English language that makes it possible to produce writing that is difficult in the extreme to understand.

Here is M.'s conclusion, which you would think he would strive to make as clear as possible:

The present analysis has shown the defectiveness of the theory on which Gödel's 1931 Theorem is based, Gödel's Theorem having validity only relative to that theory and otherwise proceeding by fallacious equivocation.

Following this conclusion there is now needed a re-appraisal of the foundations and significance of mathematics in the light of the present analysis indicating the relevance of mathematics to fundamental aspects of empirical reality via what is meant by "a value" as derived from fundamental aspects of our empirical experience. Various structural (mathematical) systems may then be defined on the basis of various systems of interrelated values, the relations between the values determining the structure of the system. Thus, starting from an initial basis derived from fundamental aspects of empirical reality mathematics may develop *hypothetical* systems (as in "pure" mathematics) or systems to correspond to and model actual systems of the empirical universe (as in "applied" mathematics) *on a basis derived from the empirical universe*. The relevance of mathematics to empirical science and to practical life thus becomes understandable, and also the activities of "pure" mathematicians can be viewed within the general context of the set of potential structural systems based on fundamental aspects of empirical reality.

In *A Budget of Paradoxes*, Augustus De Morgan included a legend explaining why the English language is the way it is. In ancient times, the story goes, the English had no language and communicated by hissing. They petitioned the devil for a language; he prepared a cauldron

> and the air was darkened by witches riding on broomsticks, bringing a couple of folios under each arm, and across each shoulder. I remember the time exactly: it was just as the council of Nice had broken up, so that they got books and papers there dog cheap; but it was a bad thing for the poor English, as these were the worst materials that entered into the cauldron.
>
> There were books from the Dalai Lama, and from China: there were books from the Hindoos, and tallies from the Caffres: there were paintings from Mexico, and rocks and hieroglyphics from Egypt: the last country supplied besides the swathings of two thousand mummies, and four-fifths the famed library of Alexandria. Bubble! bubble! toil and trouble! never was a day of more labor and anxiety; and if our good master had but flung in the Greek books at the proper time, they would have made a complete job of it.

But he skimmed off the froth too soon and distributed it to the English. The result was:

> The fact is this; it matters not who gets up to teach them, the hard words of the Greek were not sufficiently boiled, and whenever they get into a sentence, the poor peoples' brains are turned, and they know no more what the preacher is talking about, than if he harangued them in Arabic. Take my word for it if you please; but if not, when you get to England, desire the bettermost sort of people that you are acquainted with to read you an act of parliament, which of course is written in the clearest and plainest style in which anything can be written, and you will find that not one in ten will be able to make tolerable sense of it. The language would have been an excellent language, if it had not been for the council of Nice, and the words had been well boiled.

Something of M.'s should have been better boiled. He did not go into what implications the falsity of Gödel's Theorem would have.

GOLDBACH CONJECTURE, THE

Mathematical immortality strikes just as randomly as other sorts of immortality. Some mathematical immortals—Archimedes, Newton, and Gauss, for instance—merit recognition not just from students of mathematics, but from the entire human race. Others, of the caliber of Euler, Lagrange, Riemann, and Hilbert, richly deserve their place among mathematical immortals. However, some other names in the mathematical pantheon are there through luck, or because someone made a mistake. It is not fair, but life after death is no more fair than life before it.

An example of fortuitous immortality is given by Christian Goldbach (1690–1764). Goldbach was a mathematical amateur who had the advantage of being able to write letters to Euler and to have Euler answer them. In 1742, he wrote to Euler:

> I consider it not inappropriate to jot down those propositions which are very probable, notwithstanding the fact that a real demonstration is lacking. For, even if they are found later to be false, they can still give an opportunity for the discovery of a new truth.

He then mentioned his discovery that every even integer he looked at, from 4 up to a hundred or so, was a sum of two primes. Sure enough: $6 = 3 + 3, 8 = 5 + 3, 10 = 7 + 3 = 5 + 5, \ldots, 1,000,000,000 = 999,999,929 + 71, \ldots$ (Goldbach did not find that last example). He asked if Euler could find a proof. Euler could not, nor could anyone else, and so the statement "Every even integer greater than four is a sum of two odd primes" became known as the Goldbach conjecture. The Goldbach conjecture it will remain: even if Jane Doe, as a result of staggering ingenuity, hard work, and probably a little luck, proves that the conjecture is true, it will not go down in history as "Doe's Theorem." She will survive only in parentheses: what the

books will say is, "Goldbach's conjecture, proved by Doe in 2032, states that . . . ."
No matter what happens, the name of Goldbach is immortal.

It is not fair. *Anyone* can make conjectures. Conjectures are cheap. Goldbach
also made one about odd integers: every odd integer, he conjectured, is the sum of
a prime and twice a square. Just as with his other conjecture, it is true for small
integers:

$$5 = 3 + 2 \cdot 1^2, \qquad 7 = 5 + 2 \cdot 1^2, \qquad 9 = 7 + 2 \cdot 1^2, \dots,$$

but Goldbach's other conjecture fails at 5777, so his name is not attached to it. If
a route to immortality is making conjectures, perhaps if you make enough of them
you will be remembered for one. I may as well try: I conjecture that every composite
odd integer is the sum of a prime, a square, and one or two cubes, as

$$9 = 3 + 2^2 + 1^3 + 1^3, \qquad 15 = 5 + 3^2 + 1^3, \qquad 21 = 3 + 4^2 + 1^3 + 1^3.$$

If that conjecture does not do the trick, more can be supplied.

But we should not think ill of Goldbach, since he did more than make random
guesses. Though an amateur, he was a competent mathematician. In 1724 he wrote to
one of the Bernoullis proving that the product of three consecutive positive integers
is never a perfect square, something that is not trivial. Also, the reason he made
his conjecture was that he wanted to help out Euler. Euler was trying to prove that
every integer is a sum of four squares. He first considered the problem in 1730,
and he kept after it for more than forty years. At the time Goldbach wrote, Euler
had proved that every prime of the form $4n + 1$ could be written as a sum of two
squares. Thus, if Goldbach's conjecture was true and an even number happened to
be a sum of two primes of the form $4n + 1$, then that even number would be a sum
of four squares and part of Euler's problem would be solved. (Lagrange was the
first to prove that every integer is a sum of four squares, which he did in 1770 using
ideas that had nothing to do with the Goldbach conjecture. In 1773, Euler gave a
simpler proof—success at last.) So, Goldbach's heart was in the right place. He was
not seeking immortality; it was thrust upon him.

The Goldbach conjecture is true, for sure. It was verified for numbers up to 1000
in 1894, to 100,000 in 1940, to 33,000,000 in 1964 (with the help of a computer),
and to 100,000,000 in 1965, a feat duplicated in 1980. The limit could be raised
further, but there is no point. The reason is that the conjecture is so *very* true: not
only is every large even integer a sum of two primes, it is the sum of two primes
in *many* ways. There are reasons for believing that an even integer near N can be
written as a sum of two primes in approximately $N/(\ln N)^2$ ways, and numerical
evidence confirms this. Thus, for even numbers near 100,000,000 there are around
300,000 ways of writing them as a sum of two primes, and it is hardly worth the

computer time to find them. The representation of 1,000,000,000 as 999,999,929 + 71 is only one of approximately 2,300,000 such representations. There are so many primes that you do not need to use all of them. Numerical evidence shows that you can do with far fewer: if there are P primes available, you can get along with about $P^{.6}$ of them, properly selected. So, of the approximately 5,500,000 primes from 1 to 100,000,000, you can write all of the even numbers up to 100,000,000 as sums of two primes using a mere 11,000 (or so) of them. No one doubts that the Goldbach conjecture is true.

Proving it is another matter. The reason that the proof is difficult is that the primes are multiplicative objects and the Goldbach conjecture is about addition. Primes are the numbers that have no proper divisors, and every integer is the product of primes. There is nothing about addition there. When the Goldbach conjecture is finally proved, it will probably turn out to be true not only for primes but for infinitely many other sequences of integers that behave like the primes—sequences that are fairly dense among the integers and that have no unusual gaps in them. Denseness is necessary: not every integer is the sum of two squares because there are not enough squares. Also necessary is the absence of unusual gaps:

$$0, 1, 2, 3, 9, 10, 11, 12, 13, 14, 15, \ldots$$

is about as dense as a sequence of integers can get, but not every even integer is a sum of two of its members.

Efforts to prove the conjecture have advanced mathematics considerably since Euler. Although mathematicians have not succeeded in proving the conjecture, methods have been developed that are useful in other problems as well as being interesting in themselves. That is how mathematics advances, by attacking problems. In 1930, Schnirelmann showed, using density arguments, that every even integer is the sum of no more than 800,000 primes. That rather large bound has been reduced over the years, being brought down to 19 in 1980. That is still quite a way from 2, and since no improvement on 19 has been made since 1980, the limit of density methods may have been reached. In 1937, Vinogradov, using the circle method of Hardy and Littlewood, showed that every sufficiently large integer, even or odd, could be written as a sum of three primes. "Sufficiently large" was very large indeed,

$$10^{10^{10^{10^{17.86}}}}.$$

Later work has reduced that huge number somewhat—in 1989 to $e^{e^{e^{e^{11.503}}}}$, or a mere $10^{43,000}$, but that is still too large for computers to check all the smaller numbers. The circle method may also be near exhaustion. Sieve methods, descendants of the sieve of Eratosthenes, have also been used. In 1920, Brun showed that every sufficiently large integer could be written as a sum of two integers, each having at most nine prime factors. In 1923, Rademacher showed that every even integer, large or not, was a sum of two integers with at most 7 prime factors. Over time, other workers improved this: in 1940 to four prime factors for each of the two integers, in 1956 to one and four, and by 1973 to one and two. *Almost* one and one, which would be Goldbach's conjecture, but not quite. There we have stuck, so sieve methods also may have done all they can and some entirely new idea may be needed to prove the conjecture. Or, some small improvement in previous methods may be all that is required to polish it off. Time will tell.

The Goldbach conjecture has attracted cranks, though not as many as angle trisection or Fermat's Last Theorem. The reason is that an angle trisector does not have to know anything more than school geometry to solve (he thinks) the problem, and most fermatists attack FLT with no weapons more advanced than high-school algebra. The properties of prime numbers do not come up in everyone's mathematical education, nor is the content of the Goldbach conjecture something that everyone knows, so the pool from which provers of the conjecture emerge is smaller than the lake, or ocean, containing trisectors or fermatists.

Nevertheless, goldbachers exist. Here is the first half-page of a proof of the conjecture, with nothing left out:

GOLDBACH CONJECTURE

$$F(m) \leq \sum d(4m + 1)$$

$$F(m - 2) \leq \sum d(4m - 3)$$

$$F(m) - F(m - 2) \leq d(4m + 1) + d(4m - 3)$$

If m and $4m + 1$ are primes for infinitely many m

$$F(m) - F(m - 2) \leq 1 + d(4m - 3)$$

$$F(m) \approx F(m - 2) \leq \frac{\sqrt{m} \log m}{8} \text{ or } \frac{\sqrt{m}}{8 \log m}$$

If $m - 2$ is not a prime it is $\dfrac{\sqrt{m} \log m}{8}$

If $m = 4k + 1$ $F(m - 2) \approx \dfrac{\sqrt{m} \log m}{8}$

but $F(m)$ may be $\approx \dfrac{\sqrt{m} \log m}{8}$

so $(m - 2)$ must be a prime

$$F(m) \leq \frac{\sqrt{m} \log m}{8}$$

for infinitely many values.

The proof goes on similarly. The person who wrote that had extensive mathematical training. But the principles of good mathematical exposition—define what your symbols mean, give reasons for your steps, and so on—did not sink in enough. It is hopelessly incomprehensible.

V. V., another prover of the Goldbach conjecture, also had more mathematical training than the average angle trisector or fermatist. His proof drew its inspiration from the sieve of Eratosthenes. He said, look at the N possible ways that $2N$ can be split up so as to be a sum of two integers:

$$1, \quad 2N - 1$$
$$2, \quad 2N - 2$$
$$3, \quad 2N - 3$$
$$\vdots$$
$$N, \quad 2N - N.$$

We can immediately discard half of the pairs since the numbers in them are both even, cannot both be prime, and thus cannot satisfy the conjecture. There are now $N/2$ pairs left. Of those, one-third have the number in the first column divisible by three and hence can also be discarded. Another one-third have the number in the second column divisible by three, so we get rid of them too. Thus, we have eliminated two-thirds of the $N/2$ pairs, so the number left is

$$\frac{1}{3}\left(\frac{N}{2}\right)$$

Of those, V. said, one-fifth of the numbers in the first column are divisible by five, as are one-fifth in the second column, so away go two-fifths of the remaining pairs, leaving

$$\frac{3}{5} \cdot \frac{1}{3} \cdot \frac{N}{2},$$

and so on. We continue throwing out $2/p$ of the pairs remaining at each stage for $p = 7, 11, 13, \ldots, P$, where P is the largest prime less than $\sqrt{N}$. The number of pairs left is

$$\frac{P-2}{P} \cdot \ldots \cdot \frac{3}{5} \cdot \frac{1}{3} \cdot \frac{1}{2} \cdot N.$$

Since this is obviously greater than zero, the conjecture, V. said, is true.

If we apply V.'s idea to $N = 32$, the pairs are, after the even pairs have been deleted,

1	3	5	7	9	11	13	15	17	19	21	23	25	27	29	31
63	61	59	57	55	53	51	49	47	45	43	41	39	37	35	33

and crossing out only the pairs that contain a multiple of 3 or 5—the only primes less than $\sqrt{32}$—we get

3	5	11	17	23
61	59	53	47	41

exactly the five ways of dividing 64 into two prime parts.

The difficulty in applying this argument in general is that the second element in the pair is not independent of the first. Among the pairs with a prime first element, *on the average* there will be pairs with primes as second elements. But Goldbach's conjecture is not that an even integer will be a sum of two primes on the average; rather, it is that *every* even integer will be a sum of two primes. Information about what happens on the average is interesting, but is not of much use in proving theorems about what happens in every case.

What V. had done was to rediscover the information that can be derived about the Goldbach conjecture from the prime number theorem. That famous theorem says that, on the average, the proportion of prime numbers among numbers of size close to N is approximately $1/\ln N$, and the estimate gets better and better as N gets larger. Using this, the proportion of pairs $(k, 2N - k)$ that will consist of two primes is, on the average,

$$\frac{1}{\ln 2N} \cdot \frac{1}{\ln 2N}.$$

So the total number of ways of writing $2N$ as a sum of two primes will be, on the average,

$$\frac{2N}{(\ln 2N)^2}.$$

V.'s estimate is close to this because

$$\ln \prod_{3}^{N} \left(1 - \frac{2}{p}\right) = \sum_{3}^{N} \ln \left(1 - \frac{2}{p}\right) \approx \sum_{3}^{N} \left(-\frac{2}{p}\right)$$

$$\approx -2\ln(\ln N) \approx -2\ln(\ln 2N) = \ln \left(\frac{1}{(\ln 2N)^2}\right).$$

V. even realized, partly, that what he had found was just a probabilistic estimate, since in his proof he wrote:

> Therefore there are at least
>
> $$G_{2N} = \left[N \left(\frac{1}{2} \cdot \frac{1}{3} \cdot \frac{3}{5} \cdot \frac{5}{7} \cdot \frac{9}{11} \cdot \dots \cdot \frac{P(k) - 2}{P(k)} \right) \right]$$
>
> prime representations of $2N$.
>
> This equation, for all practical purposes ($2N \geq 120$) gives the lower limit for the number of prime pair representations of $2N$. When the equation is to be used for $2N \leq 120$, or when a more accurate estimate is desired, corrections must be introduced to the simplifications we made.

Of course, the keen eye of a mathematician who examined V.'s work noticed that paragraph, and the mathematician told V.:

> Such statements as the paragraph at the top of page 7 are entirely out of place. What is meant by "for all practical purposes"? (There is nothing practical about Goldbach's Conjecture.) What is special about 120?

Of course, V. got mad:

> You assumed *a priori* that I *must* be wrong and that it is a waste of time to understand clearly my argument. Thus it happened that you imputed to me an elementary error that I did not make.
>
> You therefore will not be surprised to learn that you *did not* disprove what I still hold to be the proof of Goldbach's Conjecture. To do that *at least one* error must be shown in the proof. Perhaps you do not need that to dismiss [my] paper, but you sure have to show that before you can dismiss the possibility of its existence.

One can sympathize with V., a little. It is frustrating to be told that since the probability that an amateur will prove Goldbach's conjecture is .0001 or less, it follows that your proof is almost certainly wrong. However, V. *was* wrong. His whole approach to the problem was wrong, so no matter what he said or how he arranged the details of his proof, it would be dismissed. Since he could not understand that, he could not understand why his demand "Show me one error in my proof" went unanswered. This is a very common pattern in crank mathematics.

Another worker on the conjecture, J. C., should be labeled an anti-goldbacher since he asserted that the conjecture is *false*. He said:

> There are infinitely many odd multiples of 4 which cannot be expressed as the sum of two primes. (Goldbach's Conjecture is false.)

C. also had more mathematical training than the average person. He showed mathematical talent early, he claimed:

> When I was 15 and just beginning my first formal course of study in
> Euclidean geometry, I was approached by the head of the high school math
> department and heartily congratulated for an outstanding performance in
> the preceding semester's algebra course (my first formal course in algebra). He then showed me a geometry problem which he was wondering if
> I could solve. As soon as he had finished explaining to me the particulars
> of the problem (I was being polite to wait for him to finish talking), I told
> him the answer. His jaw dropped. His eyes grew large. And when he recomposed himself he said, "That's right! How did you get it?" I explained
> as best I could but he said he didn't understand. . . .
>
> It wasn't long before I was convinced that there was something different about the way I thought about mathematics (and the world in general),
> but I couldn't find the words to nail it down. . . .
>
> After many side steps, I finally earned a BS degree in Mathematics
> from [a state university] in 1977. Immediately I began my own research
> into mathematics.

The fruits of his research were, besides the falsity of Goldbach's conjecture, a proof of Fermat's Last Theorem and a proof that there are infinitely many primes p such that $p + 2$ is also prime.

His disproof of the Goldbach conjecture was contained on a half page, but I could make no sense of it. I then told C. that, if he wanted to show that the Goldbach conjecture was false, he should give a specific example of an even integer that could not be written as a sum of two primes. I should have known what he would reply:

> I have not the knowledge to exhibit a specific even integer that can't
> be written as a sum of two primes, even though there are infinitely many
> of them. They are simply too large for me to handle.

He went on to add:

> The proof you seek is one which is based on knowledge and confirmed by knowledge. The proof I offer is based on understanding and
> confirmed by wisdom. Knowledge and understanding are held together
> by wisdom. Therefore wisdom is the principal thing, *not knowledge.*

I concluded that the special way he had of looking at mathematics and the world was a mystical way. There is no use in trying to apply rationality to mysticism.

GREED

A one-sheet advertisement from California, written by A. S. and addressed to

> All Intellectual Organizations and Individuals,

announced that

> Diligent research by SIR has brought to light history's most shocking mathematical Boo-Boo.

SIR stood for the "[S.] Institute of Research," of which S. was "Director-Solerthologist." I doubt that many people sent in the coupon at the bottom of the sheet, since the price S. was asking was high:

> Enclosed is $100.00 ($5.00 for additional sets), check, money order, traveler's check, (or U.S. stamps), made to SIR, for which you are to rush to me—Air Mail—Postpaid—a set of 12 pages of easy to understand mathematical calculations and proofs.

I suppose that there is no way to stamp out the delusion that mathematics can be sold. For an answer to someone who suffers from that delusion, I cannot improve on this quotation from a mathematician writing to an angle trisector:

> You remark that you would like to "realize all benefits which will accrue" to you under existing copyright laws. If you mean financial benefits, I hasten to assure you that there are none. No one has ever yet sold a mathematical theorem, however important. All the hundreds and thousands of mathematical papers that are published annually in the learned journals gain their authors not one cent. It is considered an honor, or at least a privilege, to be able to publish. The only financial benefits that ever might

be realized would be the indirect ones, such as ultimate promotions for sufficient publication, etc. Probably one could sell an engineering idea, if one knew how; but not a purely mathematical one, like angle trisection. The reason is simple enough: no one could use it for anything, except the furtherance of knowledge; and the furtherance of knowledge has never been a lucrative business.

Another mathematician wrote:

> *A Fable.* Once upon a time a mathematician named Pierre de Fermat proved a theorem. The manuscript was in his baggage and got lost when he was travelling. The airline gave him $50.00.
> *Moral.* Take the money and run: few of us are paid so much for our theorems.

What S. was trying to sell was proof that Newton's law of gravitation was incorrect. His level of mathematical knowledge can be judged from his statement that

> Even a mathematician—if he has been away from it a few months or years—may confuse diameter and circumference.

S. later retreated from his $100 demand and published, at lower cost to the public, a book with the catchy title *Newton's Laws Are Full of Flaws*. Its contents are physics crankery and hence beyond the scope of this book, but it contained one thing that was new to me and may some day come in handy: a kilometer is a ton less than a mile. So it is, since $5280 - 2000 = 3280$ is the number of feet in a kilometer. The greatest integer in the number of feet, actually, since S. said that the number of feet is, to one decimal place, 3280.8.

Another person who wanted a great deal of money for his work was D. I., who wrote a book in 1981 and circulated a pamphlet that is a "conspectus" of it, giving a summary of what it contains. Part of the contents follow:

> II. Mathematical Hocus-Pocus   13
> Lays open to view the circular reasoning by which mathematics makes use of the concept of subtraction in defining negative numbers and then defines subtraction as adding a negative number. Spotlights, as well, the necessity of distinguishing between the use of the minus sign to indicate subtraction and its use to indicate transposition. . . .
> IV. New Curves from Old Equations   51
> Explores the often remarkable effect worked by "eteomathematics" on various curves studied by analytic geometry. . . .
> VI. Extent as Divisibility   107

This carries on to Chapter XXI, "A Matter of Blind Faith," which begins on page 339. I would like to have had a copy of his book to see what his doubts about infinity were and how imaginary numbers can be done away with. However, the price was a bit steep: $500. The reason for this was:

> Advanced thinking typically proceeds partly in private and otherwise in consultation with a peer group of roughly one or two dozen people. For a variety of reasons not all hard to surmise the thinking embodied in this book has had to be carried on in secrecy over many years, except for a period of a few months when some overtures were made to the physics community with very interesting negative results. Having otherwise worked in dogged secrecy, the author now needs to establish contact with an appropriate peer group of fellow dissidents.
>
> To use for this purpose the information-handling apparatus of the science community would be worse than a waste of time; it would invite use of that apparatus as an instrument of suppression. In order not to attract much attention from the personnel of that apparatus a high price for copies is being established, no review copies or news notices are being sent to appropriate English-language journals, and there will be no resort to the common promotional expedients of book publishing for profit in this country.

"Overtures were made . . . with very interesting negative results." I. must have circulated earlier work and been hurt by being ignored or rejected. Once burned, twice shy: I think that I. wanted to make sure his readers would have so much invested in his work that they would not dismiss it in the way that crank work usually gets dismissed. Or perhaps I. did not want to have readers: his writing is filled with railings against the scientific establishment for its closed-mindedness and other sins. Perhaps the pleasure of being able to complain that he was ignored is what he was really after.

Big money is not to be made out of mathematics. Some people with mathematical training—very few—strike it rich by writing a best-selling textbook, but that is pedagogy and not mathematics.

INCOMPREHENSIBILITY OF CRANK'S WORKS

What went through the mind of F. S. when he produced the following, what does it mean, and why did he mail 1,500 copies of it?

DEAR PROFESSOR

I am optimistic my thoughts and ideas will be useful for peaceful means. After extensive research I came to the conclusions that Pythagoras discovered the present system of mathematics now in use about 500 B.C.

What transpired before? The pyramids in Egypt appear to be the first known works in the ancient world. The beginning of all known mathematics must have appeared then.

Two new discoveries in mathematics are contained. The first one is a *series of single progressive numbers* that extends from zero to infinity. Secondly, the remaining theory concerns the division of the number zero into fractions and quasars. The two diagrams prepared explain the principles involved very clearly.

This notice is being sent to about fifteen hundred universities, attention of the Chairman of the Mathematics Department.

single consecutive numbers

0   1   2   3   4   5   6   7   8   9 $\overline{0}$ 10   (non existent)

an infinite series of   the names are as follows
single consecutive numbers

0	1	2	3	4	5	6	7	8	9		
$\overline{0}$	$\overline{1}$	$\overline{2}$	$\overline{3}$	$\overline{4}$	$\overline{5}$	$\overline{6}$	$\overline{7}$	$\overline{8}$	$\overline{9}$		

a   aaaba — i
b   bbbcb — j
c   cccdc — k
d   ddded — l
e   eeefe — m
f   fffgf — n
g   ggghg — o
h   hhhih — p
i   iiiji — q
j   jjjkj — r

infinity

negative

$m/0$ $/0$ $h/0$ $/0$ $k/0$ $/0$ $j/0$ $/0$ $\text{inf}/0$ $/0$

positive

$\overline{8}/\text{inf}$ $/\text{inf}$ $\overline{0}/\text{inf}$ $/\text{inf}$ $3/\text{inf}$ $/\text{inf}$ $2/\text{inf}$ $/\text{inf}$ $1/\text{inf}$ $/\text{inf}$ $0/\text{inf}$ 0

infinity

a quasar (less than a fraction)

The number scale extends from negative zero to positive zero. All positive numbers in between are single fractions and they extend to infinity. All negative numbers in between are single fractions and they extend to zero. At any point something less than a fraction is called a quasar.

That is all there is. No further explanation, no promise of further information, no book for sale, nothing. As far as I know, S. sent out his fifteen hundred copies and then subsided. It is likely that he got very few replies.

Many other cranks produce work that is incomprehensible, but seldom on this scale.

INFINITY, DIFFICULTIES WITH

Infinity is so large that it can overwhelm the mind. We are finite, the world is finite, the *universe* is finite. But then we learn that there are infinitely many positive integers, and that in a tiny line segment there are infinitely many points, an infinity infinitely larger than the infinity of positive integers. The mind can be overwhelmed.

The ancient Greeks, who first considered almost all of the problems of the mind that we are still considering today, were the first to ponder some of the mysteries of the infinite. Some thinkers say that Zeno's famous paradoxes—that Achilles can never catch the tortoise (because the tortoise had a head start in the race, so whenever Achilles comes to a point where the tortoise was, the tortoise will always be further along), that arrows cannot fly (because at each point of time the arrow must be somewhere and if it is in a definite place it cannot be moving), that you in fact cannot get there from here (because you must first get halfway there, then cover half of the distance remaining, then cover half of *that* distance, and so on forever)— were really meant as arguments against infinities. They were meant to show that if you supposed that space or time was infinitely divisible you got yourself into grave logical difficulties. Zeno meant you to conclude that Parmenides's vision of the universe was correct: that it is timeless, thereby eliminating difficulties about the nature of time, and unchanging, so that questions about motion in space do not occur. There are no Parmenideans around today, so Zeno did not succeed; but his paradoxes are still discussed, and the infinite continues to boggle minds.

By the way, it is slightly surprising that there are no groups of neo-Parmenideans active today. Here is a chance for an enterprising cult leader to get started. The Parmenidean universe is immensely *comforting*. It has no infinities. Motion and change are illusions, and the universe is an eternal lump, lying there, forever the same. Contemplate the lump: does it not induce a feeling of peace? What we perceive as motion, a neo-Parmenidean would say, is similar to what happens at the movies, where our minds translate an unchanging strip of celluloid, showing us twenty-four unmoving pictures a second, into the illusion of continuous motion. When we see Achilles catching up with the tortoise, or an arrow flying through the air, we are not perceiving motion: we are experiencing successive cross sections of the immutable universe and incorrectly interpreting them as things changing. Achilles does not move, nor the tortoise, nor the arrow. Illusions all. The Church of Parmenides has a future.

Here is part of a letter written by A. S. in 1977, when Watergate was fresh in everyone's minds:

> The magnitude of corruption in the universities (and probably in the lower levels of education) is to the corruption of Watergate as are the oceans to a drop of water! We shall break this corruption into three (3) categories and deal with and name two as follows: the scientific frauds and the reputation intrigues.
>
> Let us first deal with scientific fraud and begin first with mathematics. Mathematics must rightly be regarded as man's most precise discipline. Yet it is terribly flawed, and if it is that badly flawed what can we say about the other and less precise disciplines?
>
> Let us consider a point and line in the domain of plane Euclidean Geometry. Geometry is applied to the factual world, yet there are no points here. Can someone hand me a point? *No.* Because points occupy no space. As a matter of fact he could not even hand me a molecule which does occupy space. Since lines and planes have no width they also have no existence. Therefore, plane and even solid Euclidean geometry have no existence! Then where do they exist? They exist along with most of mathematics in what I call mathematical fairyland. Most of fairyland mathematics that is applicable does have *empirical* validity. But therein lies the rub. When will empirical procedures fail? The error pile-up in high speed computers establishes that there are already cases of failure.
>
> We shall show that even in mathematical fairyland Euclidean Geometry contradicts dynamics. The result is a major explosion in the domain of all organized deductive theoretical disciplines.
>
> It is of utmost importance to reorganize these disciplines. This reorganization should be done whenever possible by axiomatization. Even Euclidean Geometry and the calculus should be axiomatized probably in terms of a new notion of equality, bounded equality, as well as the old notion of equality called logical identity. Furthermore this development

of the calculus should be made in terms of either a finite number of finite decimals or a finite number of fractions.

As a matter of fact since the proof of the contradiction of geometry and any point-dynamics makes use of the infinite it establishes the fact that the use of the infinite in mathematical fairyland must be terminated. Among others, this destroys the calculus as well as the branch of mathematics known as topology! About 98% of mathematical research being done today uses this dangerous notion of the infinite. Cantor, the founder of a consistent use of the infinite was consistent with the past but he was not logically consistent. I take a position even to the left of Kronecker, Brouwer and Weyl.

S. agreed with Zeno:

The philosopher Zeno of Elea (495–435 B.C.) created what is known as Zeno's paradox—which rightfully should be called Zeno's contradiction (unfortunately paradox is defined both as a contradiction and as an apparent contradiction). The contradiction between geometry and point dynamics is not "an apparent contradiction". If we inject greater precision in one of his several arguments that "motion is impossible", we come up with an argument similar to mine that Geometry contradicts point-dynamics. A more precise version of Zeno's proof goes as follows: Motion of a point is impossible, because for it to leave the initial point of a line segment it must reach the middle point of the segment before it reaches the end; but before it reaches the middle point it must reach the quarter point, and so on indefinitely. Note that these arguments that cause contradictions make use of the infinite and really prove that the use of the infinite in mathematical reasoning causes contradictions and must be forbidden!

The contradiction between geometry and point-dynamics comes from the fact that, from the above argument, points cannot move whereas particles of matter can.

S. had a point. There are no infinities in nature, nor any continuua (at least none that can be measured), so the neat formulas of calculus, based on infinitely divisible space and time, are mere approximations to reality, which is *grainy*. There are no Euclidean points and lines: *everything* is three-dimensional. The point we make on the paper is actually a semi-hemispherical blob of ink, lines are cylinders, and planes are thin parallelepipeds. If we are going to live our lives in fairyland then the infinite is all very well; but if we are to apply mathematics to reality, S. said, then the constructions of mathematics should not conflict with the world as it is.

He went on:

> This of course should be confusing for students but they are *brain
> washed*. Brain washing is the principal enterprise of the universities (or
> as I call them the Brain Laundries). Our old, worn, tired theories are
> used to brainwash the students with this dirty mess, without constantly
> re-examining their foundations and structures. Of course this nonsense
> filters down to the high schools and elementary schools.

That is criticism of mathematics. The line dividing critics of mathematics from
cranks is fuzzy (as S. would certainly have agreed, since he said that there are no
infinitely thin lines), but it is not so fuzzy that it is impossible to determine on what
side S. was when he continued:

> I suggest that the government make available to me, for starters, a
> three billion dollar grant to form a team of perhaps several thousand *tal-
> ented* scientists and other specialists to reconstruct our various fraudulent
> theories and to develop plans for reconstituting our universities and other
> educational institutions in such fashion that they are not predominately
> havens for hacks.

The letter was addressed to President Jimmy Carter. Critics of mathematics do not
write letters to Presidents.

> Of course getting a team of several thousand scientists would be no
> easy matter. I believe that I am paraphrasing Carlyle when I say that rep-
> utation is an accident that occasionally occurs to men of talent. There are
> some men of reputation who are inventive, not necessarily very intelligent,
> but their inventiveness is most often unimportant—they are inventive in
> games played in fairyland.

S. was not a crazed ignoramus. He was the holder of the PhD degree in mathematics and was the author (in collaboration) of some papers in mathematical logic
that are still cited. A mathematician wrote about him,

> [S.] was once a reputable mathematician. Some years ago he wrote
> that he had been fired "for being a dirty anti-relativist." My reaction to
> that was that opposing special relativity is not a sign of intelligence, and
> apparently he has become worse.

Typical is the crankiness in the following:

> The third category referred to on page 1 is the vulgar and disgusting
> politics that infests academe. From the mystique of this politics comes
> political clout that falls on some members of the faculty and of course is

common to the administration. For them it is important to see to it that faculty members "fit the mold". Thus maintaining the status quo becomes more important than the constant improvement in the foundations and extensions of knowledge with eternal protection of the gem of knowledge, namely: freedom of speech!

S. maintained his membership in the American Mathematical Society, so part of his career can be traced from the Society's membership list. At the time he wrote his letter to President Carter, he listed no employer. In the previous ten years, he had been a consultant to a defense contractor for a year, professor at an urban university for three years, affiliated with a state school in California for a year, and "Grantee, Vaughn Foundation" for three years. He continued to list no employer until his recent death.

INSANITY

Some of the communications to mathematics departments come from people who are out of their heads. You suspect as much when you read a sheet, addressed "Secretary to Department of Mathematics," that starts:

AHMOSELITE

1600 B.C. Circa

PROJECT GAS LIGHT   PURPLE 96–97

MAXIMUM $50.00 charge
ON US ARMY SURPLUS SUGGESTED

equation $40 x = y$

function SS 20 left rotary cosine

OE_2 ZETA $7R_b$ DELTA x, y, Zero, C 2 E

$+ + $ 1, 2, 3, 4, inf

$4 xk = Y$ SS = wave theory $4 xk = y$

$40 x = y$ magnitude of 60   20 = ionized molecule

magnitude of = 40   MAGNITUDE OF 60

Minus 40   plus   60

wave theory

amplitude $-40, +60$

ANNE BOLEYN CHEMICALS G.M.B.H. DUESSELDORF

It continues on for another page and a half, and the same person also produced a quantity of similar material. It is clearly the product of a disordered mind, and the return address, a hospital, confirms that conclusion.

Another person was responsible for a whole series of communications to mathematics departments, but who he was is not clear since he signed himself variously as Harrison Smathers, Oscar Holland, Chiang Hsin, Elihu Cohen, and Aesop Abdullah, among others. His location was not clear either, since he gave his return address as Hôtel Miranda, Monaco, The Afghan Existentialist Society in Kabul, or other equally unlikely places. The postmarks, however, were all Princeton, New Jersey, or Roanoke, Virginia.

What he wrote was similar to the following:

Conscious of many faults in how I've been "handled" recently, I'm writing to ask of you of the mathematics community and of the academic world to be more merciful to poor old me!

Consider the time formula:

$$\sqrt{38.44} \quad \text{Yahren}$$

If deliverance, perhaps through the good actions of members of the mathematics community, should come before I reach this age then that would be "good luck" and a good sign, otherwise my age gets close to 41 and it's not so lucky at my birthday in mid-June.

The magic polynomials:

$$s^5 - 12s^4 + 62s^3 - 155s^2 + 171s - 72 = 0$$

Emancipation Decision (at Antietam, 1862)

$$s^5 - 13s^4 + 65s^3 - 155s^2 + 174s - 72 = 0$$
$$= (s-1)(s-2)(s-3)^2(s-4)$$

standard commercial version, integral roots are illustrative of time structure.

Let me beg of you my standard petitions concerning my wishes. To avoid, (1) hospitalization in mental hospitals, (2) persecutions, and (3) loss of health, begging your sympathetic support regarding my wishes to avoid these dangers.

Very sincerely,

1991999-2997680471
(Panama-Panmunjon)

More such material from other people could be quoted, but there is no point. It is sad stuff.

Outright raving, however, is fairly rare. Almost all cranks manage to stay out of mental hospitals. They do not always manage to stay out of institutions, however. For example, the following, addressed to "Chairman, Department of Mathematics," was scrawled in pencil:

I have achieved a solution to the 4 color map problem on a plane surface, at last which cooks the problems goose. If you are interested I shall be happy, at your convenience, to communicate it to you.

There were postscripts:

P. S. how much is 2×2?
P. P. S. what is the $\sqrt{}$ of -2 seconds?

The signature was

[G. P.], Ph.D.

in care of the X. County Jail. Crazy, or not? It is sometimes hard to say.

LEGISLATING PI

Everyone knows that the legislature of the state of Indiana once passed a law setting the value of π. Well, perhaps not *everyone* knows it, but many people with only a passing acquaintance with mathematics, or with mathematical cranks, have it firmly in their heads that one of our sovereign states once tried to impose its will on a constant of nature. Everyone may not remember all the details precisely—the date is often recalled only vaguely, and it might have been the legislature of Iowa, or Idaho, or maybe Illinois—but no one has any doubt that π was the subject of legislation.

That is an error. The state of Indiana (or Iowa, or Idaho) never passed a law about the value of π. It never even *tried* to legislate the value of π. However, the idea that it did has spread widely and is as definitely fixed in the collective mind as is the knowledge that in 1492 everyone except Christopher Columbus thought the world was flat. It will probably never be dislodged, but all of us who revere reason have a sacred duty to put error down whenever we can. The battle against falsehood, delusion, and mistaken ideas will never be won, but it must continue to be fought.

Although the true story of Indiana and π has been told over and over again, it has not been told often enough to drive the error out, and that is the reason for telling it one more time. The bill that the Indiana House of Representatives passed was not one setting the value of π by law. It was one that gave the state the privilege of using the proper value of π for free. The preamble of House Bill No. 246, Indiana State Legislature, 1897, is:

> A bill for an act introducing a new mathematical truth and offered as
> a contribution to education to be used only by the state of Indiana free of

cost by paying any royalties whatever on the same, provided it is accepted by the official action of the legislature of 1897.

There was no reason not to pass the bill, since there was nothing possible to be lost. Gift horses need not have their mouths inspected. Then as now, legislators were busy people with no time to master all the details of all the bills presented to them. And when a bill contained language like

> SECTION 2. It is impossible to compute the area of a circle on the diameter as a linear unit without trespassing upon the area outside of the circle to the extent of including one-fifth more area than is contained within the circle's circumference, because the square of the diameter produces the side of a square which equals nine when the arc of ninety degrees equals eight,

even a legislator with some spare time could be excused for spending it on some other piece of pending legislation. The last section of the bill, however, was reassuring:

> SECTION 3. In further proof of the value of the author's proposed contribution, and offered as a gift to the State of Indiana, is the fact of his solutions of the trisection of the angle, duplication of the cube and quadrature of the circle having been already accepted as contributions to science by the *American Mathematical Monthly*, the leading exponent of mathematical thought in this country.

Though the bill was introduced by a legislator from Posey County, its text was no doubt written by E. G., the circle squarer, angle trisector, and cube duplicator. G. was being a little devious in his claims in Section 3 of the bill. When G.'s "Quadrature of the circle" was published in 1894 in volume 1 of the *American Mathematical Monthly*, it was not an article, but part of the "Queries and Information" section where all sorts of miscellania could be found. The quadrature had the subhead "Published by the request of the author," thus putting it in the category of unpaid advertisement. The editors of the fledgling *Monthly* probably needed material to fill space and thought that G.'s piece, less than two pages long, would serve nicely and might provide entertainment for readers. So, while G. could assert that the paper containing his quadrature had appeared in the *Monthly*, when he said that his *quadrature* had been accepted, and as a contribution to science, he was stretching the truth more than a little. (His angle trisection and cube duplication appeared in the 1895 *Monthly* and were not distinguished: to trisect an angle, G. said, trisect its chord; and to double the volume of a cube, increase its side by 26%.) In any event, House Bill 246 seemed harmless enough, and it passed with no votes against.

Bills passed by legislatures are of interest to newspapers, and a bill whose subject was mathematics would stand out. One Indianapolis daily gave a history of

π, mentioning even Lindemann, who had shown not many years before that π was transcendental. But since the paper was *Der Taglische Telegraph*, printed entirely in German, many legislators were unenlightened.

After passing the House, the bill went to the Senate for consideration. Even though senators were and remained ignorant about what the value of π was, enough fuss was made in the press that they, being politicians, sensed that the bill should not pass. The Indianapolis *Journal* said:

> Although the bill was not acted upon favorably, no one who spoke against it intimated that there was anything wrong with the theories it advances. All of the senators who spoke on the bill admitted that they were ignorant of the merits of the proposition. It was simply regarded as not being a subject for legislation.

Action on the bill by the Senate was postponed indefinitely, and none has been taken for almost one hundred years.

As is common with circle-squarers, G. had been active in trying to get his discovery recognized. Actually, "discovery" is not quite the word, since he had written earlier:

> During the first week in March, 1888, the author was supernaturally taught the exact measure of the circle.... All knowledge is revealed directly or indirectly, and the truths hereby presented are direct revelations and are due in confirmation of scriptural promises.

G. evidently had some influence with the editor of the Indianapolis *Journal*, because that paper had an editorial that included:

> Some newspapers have been airing their supposed wit over a bill introduced in the Legislature to recognize a new mathematical discovery or solution to the problem of squaring the circle, made by Dr. [G.], of Posey County. It may not be the function of a Legislature to endorse such discoveries, but the average editor will not gain much by trying to make fun of a discovery that has been endorsed by the *American Mathematical Journal*; approved by the professors of the National Astronomical Observatory of Washington, including Professor Hall, who discovered the moons of Mars; declared absolutely perfect by professors at Ann Arbor and Johns Hopkins Universities; and copyrighted as original in seven countries of Europe. The average editor is hardly well enough versed in high mathematics to attempt to down such an array of authorities as that. Dr. [G.]'s discovery is as genuine as that of Newton or Galileo, and it will endure, whether the Legislature endorses it or not.

The editor of the *Journal* was not well enough versed in high mathematics to know that G.'s quadrature was nonsensical, and he was in addition gullible enough to swallow whole all that G. must have told him. Of course, no professor of mathematics at Johns Hopkins or the University of Michigan had declared that G.'s quadrature was absolutely perfect. You might wonder how G., a man of probity and standing, could lie through his teeth like that. It could well have been that G. did not think that he was lying. Cranks find it easy indeed to believe that they have convinced authorities of their correctness. Here is evidence of that, from De Morgan's *Budget of Paradoxes*:

> An elderly man came to me to show me how the universe was created. There was one molecule, which by vibration became—Heaven knows how!—the Sun. Further vibration produced Mercury, and so on. I suspect the nebular hypothesis had got into the poor man's head by reading, in some singular mixture with what it found there. Some modifications of vibration gave heat, electricity, etc. I listened until my informant ceased to vibrate—which is always the shortest way—and then said, "Our knowledge of electric fluids is imperfect." "Sir!" said he, "I see you perceive the truth of what I said, and I will reward your attention by telling you what I seldom disclose, never, except to those who can receive my theory—the little molecule whose vibrations have given rise to our solar system is the Logos of St. John's Gospel!" He went away to Dr. Lardner, who would not go into the solar system at all—the first molecule settled the question. So hard upon poor discoverers are men of science who are not antiquaries in their subject! On leaving, he said, "Sir, Mr. De Morgan received me in quite a different way! He heard me attentively and I left him perfectly satisfied of the truth of my system." I have much reason to think that many discoverers, of all classes, believe they have convinced every one who is not peremptory to the verge of incivility.

All of G.'s efforts were in vain. His sad obituary (June 1902) appeared in his local newspaper:

> End of a Man Who Wanted to Benefit the World
>
> Dr. E. J. [G.] died at his home in Springfield Sunday, aged 77 years. He had been in feeble health for some time, and death came at the end of a long season of illness. Dr. [G.] was no ordinary man, and those meeting him never failed to be inspired by this fact. He was of distinguished appearance and came from Virginia where he received an excellent education. He has devoted the last years of his life in an endeavor to have the government recognize and include in its schools at West Point and Annapolis his method of squaring the circle. He wrote a book on his system

and it was commented on largely and received many favorable notices from professors of mathematics.

He felt that he had a great invention and wished the world to have the benefit of it. In years to come Dr. [G.]'s plan for measuring the heavens may receive the approbation which was untiringly sought by its originator.

As years went on and he saw the child of his genius still unreceived by the scientific world, he became broken with disappointment, although he never lost hope and trusted that before his end came he would see the world awakened to the greatness of his plan and taste for a moment the sweetness of success. He was doomed to disappointment, and in the peaceful confines of village life the tragedy of a fruitless ambition was enacted.

He was doomed, indeed. Let us all give thanks, loudly and daily, that we are not cranks devoting all of our energies to convincing the world that π is exactly 3.125, 22/7, or $\sqrt{10}$. The chance that our lives will be rich, delightful, and full of meaning may be small, but it is greater than the near-zero chance of the circle-squarer.

G.'s value of π has not been mentioned. The reason is that his writing was so confused and confusing that it is impossible to determine exactly what he thought π was. His *Monthly* article starts with:

> A circular area is equal to the square on a line equal to the quadrant of the circumference; . . .

This clearly says that if you take a quarter of the circumference of a circle—$\pi r/2$ if the circle has radius r—and square it, you will get the area of the circle:

$$\left(\frac{\pi r}{2}\right)^2 = \pi r^2.$$

Solving that equation for π gives $\pi = 4$. Later on G. said:

> This new measure of the circle has happily brought to light the ratio of the chord and arc of which is as 7:8.

This says, just as clearly, that (see Figure 26)

$$\frac{\sqrt{2}r}{\pi r/2} = \frac{7}{8} \qquad \text{so} \qquad \pi = \frac{16\sqrt{2}}{7} = 3.232488\ldots.$$

David Singmaster succeeded in finding *nine* different values of π implied in one place or another of G.'s bill. G. was giving the people of Indiana a lot.

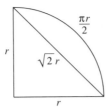

FIGURE 26

G. got the Indiana legislature to consider his value of π, but that is all. There was never an attempt by the lawmakers to set the ratio of the circumference of a circle to its diameter, to force the public schools of Indiana to teach that the value of π is 4, or perhaps 3.232488, or to enact penalties for citizens who persisted in using 3.14159.... There was only a refusal by them to take G.'s value, or values, as a gift.

LINEAR PROGRAMMING, CONSPIRACY INVOLVING

> Warning. Surprise Attack Imminent.
> is the title of a five-page pamphlet by H. V. Its subtitle is
>
> This Is No Hoax
> The American People Should Be Placed
> *On The Alert And Stay There*

And it starts:

> The American public is already under the initial subtle stages of attack. U.S. Government and private war games computers are being used against the American public to create and govern the course of crises by way of economic manipulation, to analyse the psychological crisis response of the American people using economic and other data, to calculate the best times for the next crisis, to choose the optimum type of crises, to calculate the best combination of economic controls that will yield a condition ripe for takeover of the government from the American people, and to calculate the time and method of final action that will yield the greatest probability of successful seizure of power and establishment of a dictatorship.

Standard conspiracy stuff, you may think. However, this conspiracy is unusual because it involves the mathematics of linear programming. V. knew this because

> I went to the main [M.] County library in [a city of 300,000] to obtain books on *Linear Programming*. There were very few listed in the card catalog. It is significant that their copyright dates were generally in

the 1950's and that the latest copyright date was 1961. I was able to find only two of the books on the library shelves.

A sinister and widespread conspiracy indeed, able simultaneously to prevent libraries from buying books on linear programming and to remove those that had slipped by from their shelves. Are there thousands of agents constantly monitoring nearby libraries, or are there only a few whose full-time job it is to travel and, using forged library cards, check out and never return the dangerous books?

V. described the conspirators:

> The subversive operation just described is controlled by American citizens who are themselves superwealthy, in high positions of our government, in high positions of the military, and/or in high positions in our national industries. Also included in the operation are people who will do anything for a position of political power or prestige, or for a piece of the action. The principal motivation of these people is superwealth *at the exclusion of all normal human values* and so these people constitute a mental breed or family unto themselves.

V. did not realize that people in general are such that no gigantic conspiracy can possibly remain secret, much less succeed. Some people have developed an advanced ability to see conspiracies where there are none. Companies, universities, governments: they lurch from emergency to emergency, buffeted by chance and by things they cannot control, struggling to keep their heads above water. No one has the *time* to conspire, even if the ability was there. Of course, V. would retort that that is exactly what the conspirators would like everyone to think, and that I am probably in their pay. Would that I were. I could use the extra income.

The pamphlet ends with:

> America, wake up! Big Brother is a computer and its operators, and Big Brother is watching!

MAGIC SQUARES

You would not think that a topic so seemingly harmless as magic squares could inflame the passions, but in the world of cranks passions can be easily inflamed. Magic squares are arrangements of numbers in square arrays so that the sums of the rows, columns, and two principal diagonals are all the same. Everyone knows the three-by-three square with magic sum 15:

$$
\begin{array}{ccc}
4 & 9 & 2 \\
3 & 5 & 7 \\
8 & 1 & 6
\end{array}
$$

Many people know how to construct the five-by-five square with magic sum 65:

$$
\begin{array}{ccccc}
11 & 18 & 25 & 2 & 9 \\
10 & 12 & 19 & 21 & 3 \\
4 & 6 & 13 & 20 & 22 \\
23 & 5 & 7 & 14 & 16 \\
17 & 24 & 1 & 8 & 15
\end{array}
$$

and some people could calculate that the magic sum in an n-by-n square is $n(n^2 + 1)/2$.

Magic squares are for fun, and evidence of their appeal can be found in the large volume of material that has been written about them. *Magic Squares and Cubes*, by W. S. Andrews, first published in 1908, is still in print, and that is only one of the *108* entries in W. L. Schaaf's bibliography "Early Books on Magic Squares." Magic squares have consumed vast amounts of human mental effort. For example,

A. W. Johnson did not dash off in a few minutes the following six-by-six square, with all of its entries prime numbers, that has the same sum—666, the mystical number of the Beast—in each row, column, and in each of its diagonals, whether main or broken:

3	107	5	131	109	311
7	331	193	11	83	41
103	53	71	89	151	199
113	61	97	197	167	31
367	13	173	59	17	37
73	101	127	179	139	47

Even so, not all schoolchildren have had teachers who livened their classrooms with magic squares at one time or another, nor are magic squares part of the cultural baggage carried by everyone. This is unfortunate, as H. B., writing in 1974, demonstrated. Here is his initial letter, to an English mathematician:

I have pleasure in sending you herewith a prototype version of my new game entitled Twenty One Up, in the hope that I may be favored with your highly esteemed comments, as to whether or not in your opinion, and assuming of course that such may be produced at a reasonable price, a small quantity of same may prove useful examination pieces, for testing the intelligence, integrity and sheer determination of some of our future administrators.

With an ordinary elementary education myself, which terminated at the tender age of fourteen just fifty years ago, and now appalled at the low educational standards of some of our teenagers today, I would be pleased to send you a sheet of all correct solutions to my game, on request, and now look forward to your, as already said, very highly esteemed comments with considerable interest.

The game, advertised by B. as

FASCINATING

ENTERTAINING

EDUCATIONAL

AND A

GREAT IQ TESTER

consisted in placing numbered squares on a six-by-six board so that the numbers in each row, column, and on each of the two main diagonals summed to 21. The numbers to be placed were five 1s, seven 2s, six 3s, seven 4s, five 5s, and six 6s. One solution, furnished by B., is

$$
\begin{array}{cccccc}
1 & 6 & 4 & 5 & 2 & 3 \\
2 & 3 & 1 & 6 & 5 & 4 \\
3 & 4 & 5 & 1 & 6 & 2 \\
5 & 1 & 6 & 2 & 3 & 4 \\
6 & 2 & 3 & 4 & 4 & 2 \\
4 & 5 & 2 & 3 & 1 & 6
\end{array}
$$

It is by no means the only solution. As B. wrote:

> There are of course an almost unlimited number of ways in which this may be done each giving a feeling of personal achievement and satisfaction when mastered.

You can see how even someone experienced in the ways of cranks might be tempted to reply to B. The letter was polite, and promised to send *all* solutions, which would be something of interest. An answer was sent, phrased with the characteristic straightforwardness and honesty of a mathematician:

> Thank you for the game of Twenty One Up which you sent. This is a minor variation of the very well-known idea of a magic square. Actually, the title also sounds very familiar—there is certainly the card game of Twenty-one or Pontoon. I should be interested to have your sheet of all correct solutions—I should also be very interested in your proof that you have really obtained *all* solutions....
>
> I'm afraid, finally, that I must disagree that your game will help in testing future administrators.

Straightforwardness and honesty do not always pay off. B. was angry:

> I thank you for your letter of the 3rd instant, although I find some of your statements very difficult to understand.
>
> In the first place I must emphatically refute your suggestion that my very complicated game of Twenty One Up, which can tax the ingenuity of many highly educated adults, possibly your own, is nothing more than a *minor* variation of a very simple child's game, and deeply resent such comparison.

He is right that children would be hard put to solve the puzzle. I am grown up, and it took me a little time to construct

```
2   3   4   5   1   6
5   4   3   2   6   1
4   3   2   1   6   5
2   4   5   6   1   3
6   1   5   4   3   2
2   6   2   3   4   4
```

And that solution may be only a rearrangement of B.'s, obtainable by interchanging rows and columns. What solving the puzzle implies about intelligence or administrative ability I am not sure; more than anything else it seems to test quickness in making trials.

Part of B.'s anger might have come from having his bluff ("I would be pleased to send you a sheet of all correct solutions to my game") called:

> *Secondly* I never suggested that I could supply *all* the correct solutions to my game, as there are, as clearly stated on the inside front cover of said game an almost unlimited number, but merely referred to my sheet of all correct solutions as meaning solutions without error.

Similarly, a sheet of "all correct solutions" to $x^2 + y^2 = z^2$ could contain $x = 3$, $y = 4$, $z = 5$ and nothing else. B., perhaps thinking that the best defense is a good offense, quickly moved on:

> Finally then sir, in case you may still think that my game would be of little value to your students as an IQ tester perhaps you may now care to add weight to that statement by sending me a sheet of eight all correct solutions, excluding the one I am enclosing herewith, each having been independently worked out and not one being a reflection or reversal of another in any way.

Now that might be fun, to find eight solutions not like the run-of-the-mill one above, but picturesque ones: perhaps one with a 6 in each corner, one with 1 1 1 6 6 6 down the main diagonal, one with 2 2 2 5 5 5, one with 3 3 3 4 4 4, six with a central two-by-two box with all elements the same:

```
1   1      2   2      3   3      4   4      5   5      6   6
1   1      2   2      3   3      4   4      5   5      6   6
```

and so on. Among all of the almost-unlimited number of solutions, there are almost unlimited opportunities for *tours de force*. However, busy people do not always have time for such things, and the mathematician's reply was only:

> Sorry I upset you—that was not my intention. I really would have been interested in a complete theory if you could have supplied it.

It is, unfortunately, an unalterable fact that your game is a variation on the idea of a magic square. Checking the definition will quickly show you that. (Magic squares are not childish—they have been extensively studied.)

That made B. even angrier, showing once again that you cannot often win with cranks, and that it is hard even to break even:

I hereby acknowledge the return of my Twenty One Up game and I am sure that you do not intend to be offensive—it just seems to come naturally to you.

Had you returned my game accompanied by the eight solutions requested, or even returned it without comment, as I intimated that would have been the end of the matter as far as you were concerned, but to return my letters as well, instead of filing them, or even burning them if you wished, and then go on to infer that it was I who had fallen short in some way and not yourself, I can only regard as a totally unwarranted insult.

In consequence of the foregoing then sir, it may now interest you to hear that the only Magic I am concerned with now, is the amount of Magic which it now seems that you yourself may have had to conjure up to enable you to accept my challenge, and I am afraid that I can no longer give you any assurance that this will in fact be an end to this matter.

After all sir, education is financed out of public money, and although you have now so rudely got rid of my letters, I do hope that you will not forget that I still have yours.

[H. B.]

P.S. I have rolled bigger heads than yours in the dust before today.

Whew! And people think that the mathematical life is one of calm, far from violence and free of it. Little do they know of the constant tension, the fear of bombs in packages, the apprehension when a stranger comes to the door—cranks look like everyone else, you know—and the anxiety that one's head might at any moment be rolled in the dust. Actually, cranks tend to be ferocious only in writing, and there is no record of crank-related death or dismemberment that I know of among mathematicians.

MAIL, CRANK

Some mathematical amateurs make discoveries that they think everyone should know about, so they send them to everyone. Many times this means that they sit down with a list of universities and colleges, and address a number of envelopes, from the tens to the hundreds. Of course, they want responses, but what response is not always clear. There follows an especially mysterious example: one sheet, headed

NUMBER PATTERNS

that was sent to hundreds of schools in 1974.

The sum

$$V^2 - \frac{N^2 + 1}{N}V + 1 \tag{1}$$

is zero when

$$V = N \qquad \text{or} \qquad \frac{1}{N}.$$

The minimum of the sum is $- \left(\frac{N^2 - 1}{2N} \right)^2$ at which point

$$N^2 = 2VN + 1 \tag{2}$$

is equal to zero if

$$N = V + \sqrt{V^2 - 1} \quad \text{or} \quad V - \sqrt{V^2 - 1}.$$

The minimum is $-(V^2 - 1)$.

The circle associated with (1) is obtained from

$$\csc A = \frac{N^2 + 1}{2N}. \tag{3}$$

The radius vector of the circle is $N^2 + 1$ with components

$$N^2 - 1 \quad \text{and} \quad 2N.$$

Likewise from (2)

$$\csc B = V. \tag{4}$$

The circle associated with (2) has a radius $= V$ with components 1 and $\sqrt{V^2 - 1}$.

The two circles are related by

$$\csc^2 B - 2 \csc A \csc B + 1 \tag{5}$$

as obtained from equations (4) and (3) when substituted in sum (1).

That's *it*, word for word, with nothing left out. There was no cover letter, no conclusion, no reference to any pattern of numbers, no diagrams of circles or angles, nothing but what appears above. Questions arise. One is, what was going on there? Clearly, there is a parabola

$$y = x^2 - \frac{n^2 + 1}{n} x + 1,$$

which has zeros at $1/n$ and n, and a minimum halfway in between, as in Figure 27. But where are the circles associated with (1) and (2)? No doubt they are snuggling up to the parabola somewhere; but where is that radius vector, and what are angles A and B? Angle A is in Figure 28 and angle B is probably the other angle in that triangle, but the sheet as a whole is a mystery. Perhaps it is one suitable for giving to students of mathematics: it would be a nice change for them to have a problem whose answer their teacher did not already know.

Another question is, what did the author think he was doing when he sent out his hundreds of letters? Each one was embossed with his seal, certifying him to be Registered Professional Engineer number 4631 of his state. So he was not

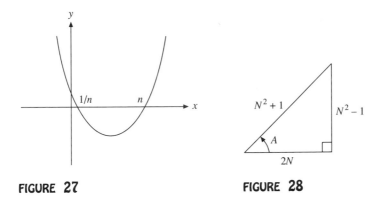

FIGURE 27   FIGURE 28

deranged or babbling, and presumably he had something in mind he thought was significant. It is another mystery. What is not a mystery is why mail such as his ends up in wastebaskets: members of mathematics departments (in whose mailboxes such mail appears) have quite enough to deal with already and have no time for nonmathematical mysteries.

MEGALOMANIA

Here we have volume 1, number 3, of *The New Renaissance*:

Published by New Renaissance Productions, Inc.

To introduce and promulgate the works of Dr. [G. K.], R. M.
America's foremost Renaissance Man,
considered to be the genius of the century,
based on his outstanding contributions
in the Arts and in the Sciences.

"R. M." stands for "Renaissance Man." The 237-page issue (containing no advertising, and all written by K.) is undated, but it appeared in late 1989 or early 1990. The headline on page 1, in letters two inches high, is

AMAZING

followed by:

[K.] Breaks Through
Solves one of Science's Unbreakable Enigmas
C = Function of Subzero decelerating to Zero
$$C^2 = A^2 + B^2, \quad C = \sqrt{2} = 1.4142\underline{135}$$
All mathematical processes are interpolations
of velocitous spirals measured in degrees.

π 22/7   [3.14285714]   &   [3.14159]   are
the result of interweaves of the other major pi's
[.714]   [3.459]   [3.777]   [4 vs]

π Prime integer 317$^+$ / 2 ["divided by PI 5"] = 159

[Because of spimes, 317 already inside #318]

$$C^3 = A^3 + B^3 \ldots C^4 = A^4 + B^4 \ldots C^N = A^N + B^N \ldots \text{ are TRUE}$$

Decimal points refer to several mixed velocities. Hypotenuse 1.41 decelerates to 3.141. Dual nature of the equal sign is exposed:

Distance [.. #] diameter based on

Velocity [..] long diameter based on hypotenusi

Obvious proof of $C^N = A^N + B^N$ lies in the combination of subzero Pi and the 01 "constant" which loses more accuracy via supervelocity $*$ megadistance. $\sqrt{-1} = 1 = \sqrt{1}$ is way off track.

Pi 3.141 is the deceleration of Pi $3.459 - .317$ as $186,000$ m/sec^2 is the deceleration of 1.86 squared, which is Pi 3.459. Famous prime π integer 317 added to $\pi\, 3.459 = \pi\, 3.777 + .317 = 4.094$ Plus an equilateral triangle (".006") = 4.1 and we start again in reverse. Hence is inside that network of "speeds of light".

$E = M^{C^2} = A^2 + B^2$: Speed of light disconstant. Theory of Relativity altered and expanded. The vacuum is the tunnel of velocity between Subzero and Zero where all numbers are made.

American Mathematical prestige soars to new heights.

As expected, Dr. [K.] clarifies, simplifies, and classifies certain mathematical phenomena that have mystified Science since the beginning of Zero. His brilliant validation of $C^3 = A^3 + B^3$ is a quantum expansion of other discoveries published here since 1987. He has proven with pristine logic that the latter equation (heretofore considered "impossible") is concrete fact at specific megadistance and supervelocity, which alters the perspective of mathematical axii.

The Planetarium Papers are leading *the greatest and most far reaching revolution in Mathematical History*. To many readers, this is no moot assumption. "I tell you," Dr. [K.] says pointedly, "... with or without humility, *subzero mathematics are re-writing the books* eclipsing several of the largest concepts in Science. *Much of it has to do with the perspectives into The Right Triangle.* And there's enough gold in my scribbled notes to compel King Solomon to choke with envy."

Perhaps. Nevertheless, this issue, as the first and second, is a blockbuster with stunning innovative steps toward *The Unified Field Equation* $(1 = 101 = 0 = 010)$ $(Pi = -1/2)$ $(\# = -Reversal\ (-92 = 29))$ $(Sub-Zero\ Pi = .714)$ $(E = M^{C^2} = A^2 + B^2)$ *(Interweaves)* *(The Decimal Point = disconstant velocity of decelerated Spime)* $(Pi = 5)$ *(The fluctuation of the Equal Sign)* *(The Long Diameter via hypotenusi not radii)* *(the cracked 01 − 10 mirror)* *(Prime integers as satellites of the 5 major*

Pi's) ($\sqrt{2}$ (1.41) as the numerical origin of Pi 3.141) (The bridges between $\sqrt{-1}$ and visible numbers after Zero). The following pages will also draw clear lines that define areas of *mathematical Chaos*.

Furthermore, Dr. [K.] divides and conquers the parameters of Albert Einstein's *Theory of Relativity* wherein the *"speed of light in a vacuum"* is contrasted and modified with the *equilateral "spime vacuum" that exists between SUBZERO and ZERO*. This perspective alters the "constancy" of the "constant = 01" whose counterpart has always been "= -1". [K.] has already proven that *numbers are decelerated forms of subzero*—negative mirrors ($-128 = 821$, $-69 = 96$, etc. (Second Issue)) Also, Einstein's *special relativity* (dealing with imaginary Time, warped Space, Gravity and accelerated mathematical; phenomena) is *altered and extended: The focus is on the deceleration of SUBZERO PI from which all numbers arise including the totally disconstant "i"* ($\sqrt{-1}$) *and its corollary mathematica natura. Slowly, surely, painstakingly and fantastically, the greatest genius in America reveals ultraviolet-infrared velocities*, causes, effects and "shunts" into numerical dimensions "that are as real as your middle finger." Dr. [K.], earthy and mirthy, DaVincian and "inchin' up to the top of tops" has established himself as one of our greatest poets and as an international Major Theorist in Mathematics. This is no surprise. As many have said: "If Nature gave him enough time and space, *[G.] could win a Nobel Prize in every category*." And this too is no revelation to those who are aware of his prowess.

All of that may not be clear but, you may think, in the 235 remaining pages the meaning will be made plain. Wrong. It is not made plain, either there or in the other eight hundred pages of the works of K. that I have seen. The other pages contain mathematics such as this from page 186:

$$\sqrt{} = 1.333\ 2938$$

$$= 1.1546834 \quad \text{"1.67+"} \quad \text{"1103"}$$

$$= 1.0745619 \quad \text{"3.459"} \quad \text{"1.7"} \quad \text{"1.41"} \quad \text{".159"}$$

$$= 1.0366107 \quad \text{"1.67"} \quad \text{"6(3)6"}$$

$$= 1.0181408 \quad \text{".714"} \quad \text{"1.86"} \quad \text{"cycle 808"}$$
$$\text{".7"} \qquad\qquad\qquad\qquad\qquad\qquad\qquad 404$$
$$\qquad\qquad\qquad\qquad\qquad\qquad\qquad\qquad\qquad 202$$

$$= \quad \text{etc.} \quad \text{"1.9"}$$

$$= \quad \text{etc.} \quad \text{"1.4"} \quad \text{"cycle 404"} \quad \text{".156–9"}$$

$$= \quad \text{etc.} \quad \text{"1.98"} \quad \text{"22/7 = 7/22"}$$

=	etc.	"3.142"	
=	1.0005619	"prime integer 317+/2"	
=	1.0002809	"1089"	"1.86–9"
=	1.0001404	".714"	"cycle 404"

THE CYCLES PROVE THAT THE C^2 of $E = MC^2$ IS NOT CONSTANT. ONLY THE NUMBERS WE USE ARE IN CONSTANT USE. WE REAL- IZE NOW THAT THOSE NUMBERS ARE DECELERATIONS OF MOST DISCONSTANT SOURCES. DEEP DOWN IN "the well" ARE CAUSES OF OUR LABELED *"CHAOS"* WHICH SPURTS SPIMES AND PRE-ZERO MATHEMATICS, FOUNDED AND GROUNDED
                    BY
(tell 'em fans 'n friends)

[K.]!!

The only thing clear there is that K. had a calculator with a square root button that could be pushed repeatedly.

K.'s use of the equals sign is one of the reasons why his work has not been accepted by mathematicians. They, not looking at things in the same way as K., tend to stop reading after encountering equations like

$$1 = 101 = 0 = 010,$$

which recurs throughout his papers, or

$$\text{PRIME} \times (1/\pi = 2\sqrt{2}) = \text{NEW PRIME},$$

also a constant theme, or

$$\pi = 50.17,$$

(*The New Renaissance*, volume 1, number 4, page 51), or

$$\sqrt{-1} = 1 = \sqrt{1}.$$

One mathematician, part of whose job it was to write letters such as the following, wrote to K.:

Thank you for sending me your materials.

I have read portions of this material and am sorry that I am not able to understand the concepts which you put forward. I can appreciate your frustration at receiving this response from me and, I expect, from others.

I wish you good fortune in your endeavor and am sorry that I cannot help you.

The letter can be quoted exactly and in full because K. included a photocopy of it in his publication. His response to it, no doubt confined to the pages of *The New Renaissance*, was:

to His Incompetence Dr. [W.]
from The Genius of the Century

His Incompetence:

I should thank you for your letter? Jesus, The Planetarium Papers are "frustrated" when mathmen of stature read "portions" of truly great manuscripts and decide in medias res what is understandable and what is not. Y'know? *You're* [deleted] [W.], on paper, voice to voice, and face to face . . . and worldwide if necessary. *You ought to be removed from your position.*

[You] have received everything I have written in Math, including the first four volumes of the Planetarium Papers. Since 1980, your softly asinine "verdict" of March '89 has been the only return correspon"dance" from your office. You don't answer the phone. *And obviously you do not recognize the supportive letters of what is, beyond question, the most powerful mathematical advance of the century.* I should take this piece of looseleaf and shove it down your throat.

What did you read? Did you start in the middle? [Deleted], you [deleted] and proudly write how you do not "understand". *Even High School students understand.* And you, with your tiny, tiny pen stuck [deleted], are telling, in proxy of thousands of mathematicians, that the Planetarium Papers are over your shoulder.

As [deleted], you "take your stand" in the middle of Modern Mathematics and "paternally" you say you are "sorry" that you, one of the "Big Boys" . . . "can't help". I should [deleted], right through [deleted]. You [six sentences deleted].

You are to send the papers to a few brilliant colleagues, my dear [deleted], you [deleted]. *The greatest pack of mathematical power just slipped through your polished nails.*

It is my undiluted *joy* to report to you the reaction of brilliant mathmen who have underlined the fact that *"incompetence", of you and others who share similar offices, is almost commonplace.* For paid personnel, writing and complaining, is practically "useless" and politically "dangerous" in terms of membership and-or publication. But since I'm so great in so many fields, you stupid little [deleted], I don't have to worry about nasty repercussions from you or people like you. If and when I ever sell my integrity, I will join your ranks and salute you with all protocol-humility and fawning without yawning. But this is not the time, Stupid. It's time "to take you out", you [deleted]. Your philosophy is like the rotten weed that losers smoke. And don't forget: this letter stands *forever*.

And His Incompetence wishes me "Good Fortune". *Good Fortune*? You must be kidding. Good Fortune. What kind of cliché is that? "I prithee thee godspeed, ye-ye, Forsooth Anon and Yonder, on and on." You little sippy sloppy idiot, *you studless cufflink*. I should wrap you in linen and send you to a Calcutta export-import. And somehow, "people" send men like you and [a mathematician] to Cambridge to "inspect" Ramanujan's papers. *I can't stop laughing. If you had read what I gave you, you wouldn't have such a "problem" trying to "decipher" $1/Pi = 2 \times \sqrt{2}$, which I discovered* on another road, ten years ago. And I gave it to you free, with hearts and flowers, issue after issue . . . for nothing, *asking only a receipt* of token gratitude "A thank you card" and notice somewhere in your publication. *You should be slaughtered, [W.]. And the one who follows you, in that office, had better fulfill my very humble request. To pay attention to the genius of the century, is the least of your duties* inside your bureaucratic bun-bungled base-brain. Ta-*da*. I don't wish you "good fortune", [W.]. *I wish you failure. I have done things in Mathematics, and many people know this, say so, and are disgusted at longtime avoidance. Geniuses and-or works of genius, die young because of men like you, [W.]. Men like you, are the punks of power. And with this little pen, I'm carving your incompetence into Mathematical History. I should* [deleted], *just to watch you walk along the squirmy line. And . . . myself and geniuses living or dead . . . don't* [deleted] *. . . if you live or die. And I speak with most lucid light.*

(signed)

[G. K.]

A different sort of letter caused K. to react differently:

Dear Dr. [K.]

I am currently a student of Mathematics and Physics at [W.] University. I have accelerated beyond the normal realms of my classmates and am finding myself sick and tired of standard conventions. The professors seem to be intelligent, but they will not recognize certain truths. They are all too afraid. They do not admit that they do not understand, and in doing so they will never find the real truth. They wallow in the security of old fashioned text books and thoughts of yesterday's minds. I am sick of it. If they had any confidence in their mental capabilities they would take a chance. They would go beyond where the great minds have left them and tackle the unknown head on. But they don't, they're all wimps. Yes, a square is not round, but is it that which makes it a square? Does mathematics change when viewed from a reference frame other than a rest frame? Perhaps mathematicians should get their minds out of a rest frame. They pass off any number when divided by zero as undefined and then just forget about it, ignoring its reality. Is that valid? I don't accept it.

Then I came across something incredible, an issue of THE NEW RENAISSANCE. I am thrilled that there is a person who is not afraid to go straight after the truth, Dr. [K.], America's foremost Renaissance Man. Right On! I realize that you must be very busy, and that I do not have any impressive title, but I am extremely interested in your *Planetarium Papers* and many of your other ideas. Any information which you could possibly send me I would greatly appreciate. I am very curious, and very excited. I am also not afraid.

(signed)

[S. D.]

The letter is carefully crafted. At first glance, it could be genuine: admissions officers, even those at a university as prestigious as [W.], cannot screen out all those prospective students who seem normal on paper but have some unseen weirdness. But it is not sincere: I learned later from someone who knew him that D. was jesting. D. wanted K. to think he was serious, he succeeded, and he had me taken in as well. I predict a brilliant future for D., and I am sure that he will in due course outgrow the callousness of youth.

K. reproduced the letter more than once, commenting in one place

THIS LETTER TO BE SENT AROUND THE WORLD

and

> The hand that wrote this letter is a perfect example of the nobility of mathematics . . . seizing the truth of discovery. Be wise to follow, lest we lose the spark of which greatness is made.

Irony can sometimes be tough to spot. Consider this other letter to K.:

Dear Dr. [K.]

I am so excited! This morning I happened to come across the most remarkable discovery of my recent 5 years, maybe my entire life! I found among our library an article in *The Planetarium Papers*. As a number theorist and rocket scientist, I have for the past 17 years had a tremendous feeling of dread . . . coming. I now realize, from a hidden awareness that something was wrong with the ordinary picture of our calculation "foundation". And then to my great delight I saw your statement:

> "AOB circumference moves as SUBZERO PI, and its velocity ($\sqrt{-1}$ obviously) is slowed down ($\sqrt{2}$) by the denser ZERO and the weight of its numbers." ... et cetera.

My understanding of this and other "meager" insights impelled me to continue and try to understand your work better. If $\sqrt{2}$ is the linear form of Zero, your revolution has begun. SUBZERO PI is an amazing "thing" to say the least.

Dr. [K.], these people around me, with whom I do my work, are well-intentioned, and intelligent *BUT* they are so closed to the other possibilities! What good is intelligence if they cannot open up their *MINDS* to a genuine breakthrough in depth, conceptions of pure Mathematics! ...

> (signed)
> [H. Z.]

Is that obviously another put-on? K. didn't think so:

> This kind of letter is "the only thing" I continue for. Can you imagine if [Z.] and [D.] "teamed up" with a few strong brilliant fearless supporters? Man, *WHAT A TEAM*!!!

Z.'s irony seems to be heavier than D.'s, but there may be none there at all. The letter was sent from "The Institute for Numerical Science and Rocketry," and in a later letter, Z. wrote:

> The Institute has been VERY busy preparing a bid for a solar wind kinetic energy transducer prototype for the Manned Mars Mission.

If Z. was trying to be funny, that may be only more funny business; but it is not clear why he should include it in his letter to K. It is possible that Z. was being completely sincere, and that he was as cranky as K., or even crankier.

It is remarkable that cranks as wild as K. are able to get encouragement. K. thought that he deserved a Nobel Prize in mathematics and he persuaded people to write to Stockholm saying that they thought so too. The Nobel Prize committee was probably able to dismiss letters from old friends like F. F.:

> [G. K.] has been a friend of mine since 1957. ...
> [G.] has been mostly successful at everything he has attempted; from a Doctorate in Languages in his early twenties to the recording of a record album for RCA; from the execution of dozens of imaginative paintings to the translation of the Holy Bible into rhyme. All of the above were

remarkable. Nothing he has done however, prepared me for the impact of the "Planetarium Papers". While not a mathematician, I. . . .

I have seen [G.] maligned, ignored and worse. The deeper he penetrates into the mysteries of Mathematics, the more "unsolvable" problems he conquers, the more perceptive his observations, the greater the crescendo of adverse reaction. This has been occurring for a dozen years or more.

In my judgment, the time has come to "Do The Right Thing" and give [G.] the recognition he is due.

On the other hand, a letter from a professor at a high-powered university, who is himself so high-powered as to have his own stationery with his impressive title "[G.] Professor of the History of Technology" printed on it, cannot be so easily dismissed and at least must be read:

I understand that Dr. [R. K.] has written you recommending Dr. [G. K.] for a Nobel Prize in Mathematics.

Since I am not myself a mathematician—indeed I must confess that I cannot fully comprehend [K.]'s work—I am not in a position to point to the particular advances or discoveries that he has made in that field. However, I am impressed by the fact that others view Dr. [K.] as a man of exceptional brilliance and his work worthy of a Nobel Prize.

Members of the Royal Swedish Academy are certainly able to evaluate the nature and worth of [K.]'s contribution to mathematics. Although some critics claim that Nobel Prizes are sometimes awarded on a "political" basis, I realize that the deliberations of your august body are above such considerations and that you indeed present these prestigious awards on the basis of actual contributions to knowledge. Therefore I am sure that you will carefully examine [K.]'s work and its impact on mathematical science in selecting the Nobel laureate in mathematics.

That is less than a ringing endorsement of K. and is as carefully crafted as D.'s letter, which is no less than should be expected of a G. Professor. It was wise of him to admit that he could not fully understand material like the following:

$\sqrt{-1}$ = "$-$" [maximus] is so fast, even Sonny Liston didn't see "it" when he dove into the canvas and signed his death-warrant. *"i" is designated as "invisible numerical velocity" (G. K.) whose bridge between supralogical mind and logical brain is $\sqrt{2}$.*

$\sqrt{2}$ is zero.

By now, you know this.

$$1 = 1^+01^-$$

.022082 is not a "mystery π"; it's minced π. Tell Grandma. *And giver the juice.*

$$.022082 = \pi = 1 = 101 = 0 = 010$$

Dear "Critic":

Do you know what's going on or are you still [deleted]? "9 lives". (Yuk-yuk). Don't laugh. (*I'll laugh. You listen.*)

That is a typical page of K.'s mathematical work that the G. Professor wanted the Nobel committee to examine carefully.

Encouragement inflamed K.'s megalomania:

You're dealing with the genius of the century, no matter how my words "hit you". You a mathematician? Mathematics needs me more than I need Mathematics.... You want to keep the edge on Outer Space and Inner Space, you need the Planetarium Papers. You can't do it without my information and its renaissance. You are to be grateful and rewarding. I wear five stars on my shoulders. They don't come easy and they're not chips to be knocked off. They're to be polished. Is that arrogance? Call it what you will. Is the light too strong for you? Wear shades or move into a shady place.... Look: suppose Picasso or Dali won the international chess championship and the heavyweight boxing championship, the Nobel prize in literature, the Fields prize in mathematics, and hit 62 home runs. This is the kind of man you're dealing with.... Those who know, know; those who don't, don't; and those who know smile on those who don't. [K.] is a megagenius. Take it for face value.

Forty pages further on:

I am composed of dedication and discipline. I am perhaps the greatest genius that America has ever had, and without doubt the greatest mathematical soldier that walks the earth today. Appreciation and public respect, are long overdue.... *I am Mathematics.* Today and Tomorrow. You do not have a "choice", as a matter of fact. Negative opinions of my personality behind this desk mean nothing to me, nothing whatsoever. Negative reactions to the revolutionary discoveries are also stock for laughter.

It is likely that K. gets few letters that encourage him. More are likely to resemble the following, from C. M., a mathematician who got his name in the newspapers for solving a noteworthy problem. K. read about it, found M.'s address, and sent him some material. M. replied:

Dear Dr. [K.]:

Thank you for sending me the 245 pages. I have tried to understand them, without success. Since I am usually able to understand mathematical writings, I am forced to the opinion that they are not mathematics. Art, perhaps, with components of humor (which I did enjoy, thank you), invective and obscenity.

But they don't give Nobel prizes for those either.

I offer the following explanation of why people do not respond to your mailings. I think it is because they believe that you will not understand them, since clearly you do not understand what mathematics is.

When you are a megagenius and get such a letter, what you do is write eleven pages of abuse, delusion, and threats, such as:

Are you . . . uh, "balanced" or what? Your reaction mocks the obvious sparkling reception of *The Planetarium Papers* by Cambridge U.K., The London Mathematics Society and the Astronomy Center at Hawaii-Hilo, etc etc et cet ter-*ah*. . . . Anyone who understands High School Mathematics cannot avoid the clarity and significance of SUBZERO PI and the relationship between hypotenuse "1.41" and Zero Pi "3.14" and hundreds of other examples proving

$$E = M^{C^2} = A^2 + B^2$$

and only [deleted] can't see the greatness of bridging the square root of negative numbers with the whole numbers via reversal #821 = −128. This is the blueprint of the mathematica neurologica. You must've got nervous. Try shiatsu. Maybe your firm will pay for it. . . .

You're just a baby. I should spank you. But I'm not gonna do that. Know what I'm gonna do? I'm going to [deleted] in every single issue, from the next, the third, until I die. . . . My advice is to re-read everything I sent you, with a clear unbiased mind, instead of wasting my precious time with your *stuttering idiocy*. Know what you are, really?

Target Practice.

. . . You're such a smart guy? Hey, pal, send the material back to me. *Everything*. . . . You sight 245 pages of gold and the only thing you can "praise" is the brown bag it was delivered in. *You're a* [deleted], *Mr. [C. M.]* You secure in your job? Hold that thought.

Listen to me, you [deleted]: it is my mission to defend geniuses and promote American mathematical progress and the *quantum leaps I have taken all alone*. It is *you* who has to watch his words. Your off-the-soiled-cuff remarks attack not only the information you've gulped without a taste, but those who have put their reputations on the line to support my supreme advances, for all of us. *You'd better send the apology of your life.* Because if you don't, I will do what I can, sooner or later, to put you in a different place.

K. had more success with one of his other projects, translating the entire Bible into rhymed couplets. Genesis begins:

In the darkness of the God
There was nothingness and Odd
There was beingness and Void
There was seeingness annoyed.

Later there is dialogue between Eve and the Serpent:

S: I live here in Eden too.
E: You do?
S: Please relax.
    I am here to tell you facts.

A story about K.'s rhymed Bible made it into a local newspaper (with a headline of "He's Psalm Kind of Guy") and even, such is the power of self-promotion, onto the first page of the second section of *The Wall Street Journal*, which headed the story

The Bible's Found a Special Niche,
But Will It Make the Author Rich?

For his Bible and other poetic works, K. was nominated by his friends for the post of poet laureate, but he was not selected.

One of K.'s former jobs was chief writer for a specialty magazine, so his pen is more skilled than the average. He attempts more in his writing than does the average crank, and the following excerpt is unique in my experience: a view, from inside a crank's head, of what is inside the head of someone who does not agree with him.

But why study the Planetarium Papers when I'm not being paid for it. I'm too busy; and this renegade wants praise and money. Well [deleted] *him*, him and his slutmouth prose. Some people think it's funny, but it's not too funny to *ME*. Yeah and some people say that he's the greatest writer in America. Ha! Can't believe that. He's not so great to *ME*, *ME-ME*, *ME* figaro figaro. Little bastid strutting around like some conquering hero, putting his greasy stamp on a thousand worlds. Who the hell is

HE . . . compared to *ME*! Him 'n his tasteless jokes and his personal "anec-
dotes". Who gives [deleted] about *his* experience! Who does he think he
is, Alexander? Alexander was *great*! Now *that* was a man. But (ooof!)
this little hump-swatter, café squatter, [city] rat, dropping his remains on
every corner. Typical [city] "vegetation", him with his "Jewish wife," Jewish dentist, Jewish lawyer, Jewish friends, Jewish psychiatrist, Jewish this,
Jewish that, and with his Arab cousins, and those Italian pals, laughing
and eating and taking pictures, kissing this, kissing that. My God! Him
and that Eddie, little Mick, *guitars in the street*! So ridiculous. And his
black baboon friends, whores and pimps I know it. Lookit him kissing her
hand! In full view! What a clown. Him and that Spic cook in the restaurant. Only the Devil knows what they're cooking up, in mid afternoon.
Little jerk should be working like the rest of us. Jesus why the hell doesn't
someone *STOP* him?! I don't have to tolerate his arrogance, dog as he is,
putting himself with Shakespeare, Dante, Einstein, Leonardo, Benjamin
Franklin . . . for God's sake! What a nut. A foodstamper no less. [Deleted]
beggar. Envelope stuffer. Can't believe that "Bible thing". Are the publishers knocking on his door? No. So it's obvious. Can't be so "great". So
he sits there so content? With that tiny little computer, pushing those little
buttons, day and night. Like a little weasel. Hypocrite. Smoking, drink-
ing. Oh isn't it nice he wears a rose in his jacket pocket. Who's kidding
whom? Come on! Mathematical madman. Humph, a "poet". Some poet.
Pencils and cheap pens, those wrinkled shirts and that vinyl bag. That's
class? Lookit this xeroxed notice: "Playwrite (misspelled naturally) Poet,
Painter, 'Entertains' on Monday Night." Sings his own songs. Yeah, he
doesn't want to sing anyone else's. Greedy [deleted]. Yeah well if I had
the time, I'd paint too, better, and build my own Renaissance, my own
clean greatness. So he's not so "unique". Good Lord's gonna teach *him* a
lesson, I tell you. "Big Genius". What a ripoff!

Let me finish with two passages from K.: one in which he summarizes his
mathematics, and the other in which he summarizes himself. The first:

All numbers are forms of "i". To get to "i" is to go below Zero. Zero
is a Pi and it has its own zero: SUBZERO PI: "i" as the circumference, $\sqrt{2}$
as the diameter. "i" is the deceleration of Pi 5 the deceleration of "∞".
Once "i" has come to "rest", the square root of two is "within rational
vision" (Pythogras' Triangle).

As $\sqrt{2}$ decelerates from "i", Square Root of Two "Minus," decelerates to Zero. Then "numbers begin". Pythogras' Triangle and $c^2 = a^2 + b^2$

is, in reality, a rational interpretation of what happens to "i" before it decelerates to $\sqrt{2}$ and "just when" $\sqrt{2}$ "creates" Zero.

In a nutshell, that's what's going on.

The second:

I am your teacher, I am your preacher, I am your hero. I built and installed a drawbridge from the island of "i" to the mainland of numbers. I did it alone. I had no colleagues patting me on the back and I didn't have a seat privvy to a patent office. There is no comparison between me and Einstein and there's no comparison between me and Newton. God bless them, but they don't take a candle to me in terms of raw mental power. Just read my resume and read everything else I've ever written. I am the greatest genius on earth. And you have nothing to say about it, save "Yes Sir". If you open your mouth, there'd better be some praise or some pleasant words. Or keep your mouth shut. If your words come at me wrapped in a hostile or martial posture, so help me God I'll cut you in half so [deleted], I'll send you back both beaten and broken.

MONEY TO BE MADE IN MATHEMATICS, LACK OF

J. H. was a hardworking and versatile lover of mathematics:

> Having little to do, I've been doing a lot of math research. I am working on—Triangles with 3 rational sides and square area. Pairs of triangles with same perimeter and area and 3 rational sides. Pairs of triangles with 2 identical rational sides, third side rational, same rational area. Magic rectangles with 4 diagonals. Magic squares with all numbers squares or cubes. Arithmetic series of 5 numbers with all of them squares or cubes. Found another solution besides $41/12$ as to what number squared plus or minus 5 gives squares (it is in the millions). Found three solutions to a box with 7 dimensions rational. Found 500 ways to have 100 coins equal to \$5.00.

Not only did he find another solution besides $x = 41/12$ to

$$x^2 + 5 = y^2 \qquad \text{and} \qquad x^2 - 5 = z^2,$$

he found that

> you can replace 5 with 6, 7, 15, 30, 84, ... but not with 1, 2, 3, 4, 8, 9, 10, 12, 16, 18, 20. It has something to do with areas of rational right triangles.

He made up new problems:

> Find longest possible series: rules—if [a term] is not divisible by 7 or 11 you multiply it by 3 and add 1. If you can divide by 7 or 11, you may ([you are not] required to do so) and subtract 1. If divisible by 77, you

can take [your] choice. Two consecutive numbers cannot be over 10,000. Chess rules of 3 times to same position and out are used.

Related is this (limit again two numbers over 10,000): find series where next number is sum of divisors of number before—if [the number] is prime you may square it for the next number.

The first problem is a variation, perhaps independently arrived at, on the Collatz problem of determining whether or not the sequences

$$a_{n+1} = \frac{a_n}{2}, \quad a_n \text{ even}, \quad a_{n+1} = 3a_n + 1, \quad a_n \text{ odd}$$

diverge or not. Since H.'s variation allows choices, it is more fun:

$$5, 16, 49, 6, 19, 58, 175, 24, 76, 229, 688, 294—$$

can you beat the sequence that starts that way?

H. even claimed to have

discovered a couple of statistics procedures, as finding a way to correct for ties in rank correlation that anyone can use with 1.6 times the trouble of no correction for ties. (The only other method, by Kendall of London U. requires much higher math skills and more than double the time.)

He also discovered a pair of odd amicable numbers,

$$69615 = 3^2 \cdot 5 \cdot 7 \cdot 13 \cdot 17 \quad \text{and} \quad 87633 = 3^2 \cdot 7 \cdot 13 \cdot 107,$$

quite correct since, $\sigma(n)$ denoting the sum of the positive divisors of n,

$$\sigma(69615) = \sigma(87633) = 157248 = 69615 + 87633,$$

even though he said that he was unable to check them. H. was a person of ability. H. had two troubles. The first was that no one would listen to him.

On January 6 you posted a letter to me containing some suggestions and a list of names. I wrote to every one on it and you again. I have not gotten a single reply in the 7 weeks that followed.

Also,

Other ideas like a list of rational 120° triangles and two odd amicable numbers seem never to get replied to from mathematics departments of colleges—but maybe I will do better in September than in August.

He constantly made comments such as "There's no interest," "Finding extremely little interest," "I find little interest," and "I find problems of this sort fascinating—but no one else seems to."

> I sent 50 postal cards to the offices of 50 large companies. 49 made no reply at all—showed no interest.

His second trouble was that he had gotten the idea from somewhere that there was money to be made from mathematical discoveries:

> It seems that to sell a magic square you have to have some sort of story. I have none—unable to sell a 26 by 26 magic square with all primes.

His rank correlation correction "seems unsalable":

> I soon learned that the only magazines who would take such an article not only won't pay, but expect me as author to buy several copies of the entire magazine containing my article.

He has had unfortunate experiences:

> I was thrown out of one meeting because I kept asking questions about ca$h. Every one there except me had a steady income.
>
> It is true that many professionals do some work for which they are not paid, but I seem to be continually asked not to be paid for anything I am allowed to do.
>
> Everyone else gets paid for his talent. Why shouldn't I?

He asked:

> Is there any money in math?—and any problem which solving gives one some recognition? I don't have it now.

Someone should have told him early in his mathematical life that there is no money in mathematics. If anyone did, he didn't listen:

> Dr. [C. T.] is a great lecturer and can answer just about any question in math save one—how can someone whose name is NOT Dr. [T.] schedule and make money on the lecture circuit?

H. felt ill-used:

> I just read a story how some mathematician at the age of 17 figured out a way to construct a 17-sided polygon and from then on had it MADE. By this I mean he had no worries from then on where his next meal, sex, or recognition came from.
>
> I am twice that age and have a lot of worries.

He was disillusioned:

> If knowing math is all that the math magazines say it is, why is it that a 99.6th percentile mathematician has made far more money as a 35th percentile typist than in math?

Generalized bitterness had set in:

> Most men of importance claim that they will listen to any point of view that "makes sense" but they don't like anyone who barges in.
>
> The trouble is that 99% have no choice. You either must be completely and totally QUIET for a lifetime or BARGE IN. . . .
>
> Anyone who has been to a legislative hearing or fought a traffic court knows that there is not a semblance of equality or justice in any court or any hearing. The big shots run it and only by hiring a big shot do ordinary citizens have a chance. . . . Nearly all major decisions—all in business, finance, politics—are made in smoke-filled rooms I am barred from (I couldn't get within a million miles of any important Democrat during the convention here, yet I as a Democrat was asked to give money to pay guards to keep me out).

H. was irredeemably ruined; so the question is, can potential future H.'s be diverted from similar downward paths? Probably not all of them, but I think a major step in the right direction would be taken if mathematics departments would answer their mail. Being ignored is never any fun, and I think it is a significant contributor to the making and exacerbating of cranks. Members of Congress answer every single letter their constituents send, no matter how cranky, and mathematics departments should emulate them. Mathematics departments in state-supported institutions have a *duty* to reply to citizens of their state. Of course one reply may generate more correspondence, but sooner or later it will cease. Even cranks can be reasonable, and even cranks get tired.

NINES, CASTING OUT

In the days before spreadsheets, before pocket calculators, even before desk cal-culators (how many remember those large, expensive, and often-broken machines made by Marchant, Friden, or Monroe that had a dozen columns of buttons labeled 0, 1, 2, 3, 4, 5, 6, 7, 8, and 9 that whirringly and errorlessly did multiplications and divisions?), in the days long ago when calculations were done by hand and had to be checked by hand, the method of casting out nines was taught and learned as a time-saving method of finding errors. It was easy to learn. To cast the nines out of an integer, say 314159, you sum its digits,

$$3 + 1 + 4 + 1 + 5 + 9 = 23,$$

and repeat as often as necessary to get a single digit. Here, one more step is all that is needed: $2 + 3 = 5$.

With a very little practice it is possible to become very proficient at casting out nines. One reason is that nines can be ignored. Casting out nines from 2399 is the same as casting out the nines from 23: $2 + 3 + 9 + 9 = 23$. Similarly, any collection of digits that sum to nine can be ignored. A skilled nine-caster confronted with 314159 would immediately ignore the 9, immediately see the 5 and 4 that sum to 9 and ignore them, immediately sum the remaining 3, 1, and 1, and immediately announce the answer, 5. Casting out nines is fun, of a sort.

What casting out nines amounts to, mathematically, is finding what an integer is congruent to modulo 9. The result of any arithmetical operation on numbers must agree with the result of the same operation modulo 9, so if you cast out nines from both sides of a true equality you get a true equality. Thus, if you cast out nines from both sides of an equality and get two different numbers, you can conclude either that the equality you started with was wrong or that you made a mistake in casting out nines. Since casting out nines is so easy (and so easy to do a second time if necessary), the conclusion is that the equality you started with is wrong. For example, if someone tells you that

$$3141592653589 + 2718281845904 = 5859874498493,$$

you quickly cast out nines from the two summands, getting 7 and 5. Thus the sum, when nines have been cast out of it, must be 3. But it is 2. You check, perhaps casting out nines from right to left instead of from left to right: still 7, 5, and 2. Thus the addition must have a mistake in it. You then give it back to the clerk who did it, saying "This is wrong." The clerk, amazed, does the sum again and this time gets it right: 5859874499493. Is

$$(31415926)(27182818) = 853973397759468$$

correct? Certainly not! Cast out nines from the left-hand side and you have (4)(1); cast out nines from the right-hand side and you have 3.

There are two troubles with casting out nines. The first is that if it shows there is an error, it gives no indication of where the error is. To find that the product above should end with 8759468 instead of 7759468, you have to do the multiplication over again. The second is that casting out nines will not catch every error. For example, as far as casting out nines is concerned, $15 + 31 = 82$ is perfectly fine, since after nines are cast out it says that $6 + 4 \equiv 1 \pmod 9$, which is true. In particular, casting out nines will not turn up the common error of transposing digits. Casting out nines may be helpful, but to be sure that your calculations are correct (even when done by a computer), the best thing is to do them over again. And even then there may be error.

The disappearance of casting out nines as a topic in the elementary school curriculum explains, partly, the publication in Singapore of

INSTANT
MATHEMATICS
For
Office and Commercial Applications

by F. S. This is a 124-page book that, S. said, was doing well:

My book entitled, "Instant Mathematics" was published in December 1988 and was launched on 10th January 1989. The first English edition of 6,000 copies was distributed throughout Malaysia and Singapore within

2 months. The second edition of 6,000 copies is in the process of printing. The books were selling so well that many major booksellers have
allocated the "Instant Mathematics" in the "Best Sellers" shelves. The
Bahasa Malaysia (Malay Language) version and the Chinese version of
the Instant Mathematics shall be ready for distribution in March 1989.

S. said in his introduction:

> In the year 1959 when I was still a senior middle student at Catholic
> High School, Singapore, I was told by one of my school-mates that one
> of the teachers was admitted to the Asylum. The cause of his illness
> was "Mathematics". I was told that he had been trying to work out a
> mathematical system which would provide immediate answers to any
> mathematical problem!
> The price he paid for his research was "insanity".

Many mathematicians would disagree with S.'s diagnosis of the cause. Be that as it
may,

> That incident sparked off my interest and curiosity. From that time
> onwards I was playing around with figures. I was having the understanding
> that the only way to get instant answers would be to work with "small and
> simple numbers", because it is only "small and simple numbers" that our
> minds are capable of perceiving easily and vividly.
> In 1962, I managed to reduce all numbers to their most basic units,
> the "one-digit numbers". In 1963, I discovered that in any calculation,
> the "one-digit numbers" of the numbers involved in that particular cal-
> culation remain the same before and after that calculation. I realized that
> there is a definite law behind each calculation, "The Law of Conserva-
> tion of Numbers". Based on this "Law of Conservation of Numbers" I
> began to develop a new system in mathematics to check mathematical
> calculations. In the process of developing this system, I discovered the
> typical characteristic of "9" and the "multiples of 9". I found that any
> "one-digit number" added to "9" or the "multiples of 9" repeats the same
> "one-digit number". The check for calculations was therefore shortened
> and simplified by the "elimination of 9 and/or the multiples of 9".

That is, S. rediscovered the process of casting out nines. But not entirely:

> As early as in 1964 and 1965 I began teaching my students to apply
> the "Instant Mathematics" in their objective questions. My students were
> taught the applications of the Instant Mathematics but no one knew how

the Instant Mathematics was being derived. I could not pronounce the Law of Conservation of Numbers then because the Law was not perfected. Though the "9" and "multiples of 9" shortened and simplified the time and checking processes, the same "9" and "multiples of 9" gave me 24 puzzling and restless years. For the last 24 years, I was struggling with the problem of "9" and the "multiples of 9".

On 1st May 1987, I managed to overcome the problem of "9" and the "multiples of 9" in multiplications, divisions, and fractions. With this accomplishment, the Law of Conservation of Numbers was perfected with 100% effectiveness.

S.'s book consists entirely of a long explanation of how to cast out nines, followed by example after example of the process. It is hard to imagine filling 124 pages with so little content, but S. managed it.

In Section V, "Applications in the Daily Life," S. wrote:

It is not uncommon to hear complaints such as, "Oh, I bought 6 items at the supermarket and I only had 5 items when I unpacked my shopping bag at home. My hair shampoo was not there!" or "The price of the butter is $2.40 but the cashier had punched $24.00" or "I bought only 12 things, however I was charged on the print-out slip for 13 things!" etc.

It is too unfortunate that most of these mistakes are only discovered after we had got home. Sometimes for the sake of one or two items, or when the wrongly charged items are not that much different in price, we might consider it too troublesome and too time consuming to go back to the supermarket to rectify the matter.

At times, we feel we are being cheated! Sometimes we might feel a bit uneasy for paying extra! What to do—all these could happen because we are careless and we don't bother to spend some time to check the print-out slip.

To prevent all these regrets and unhappy feelings, it is always advisable to check through the print-out slip before making any payments.

Checking by casting out nines, of course, after having verified that the number of items you are being charged for is the same as the number of items in your shopping basket. People in the United States do not often check their supermarket cash-register receipts item by item while the people behind them in line fume, but perhaps things are different in Singapore. They must be, because if you are not quick enough to check that the proper prices are being entered into the cash register, S. said, you can always counter-check the prices together with the cashier. She will read out to you the printed prices one by one against the items you have purchased.

S. never gave any indication that a calculation that checks by casting out nines could be incorrect. His examples all end with "addition correct," "product correct," "therefore, the new value $16.80 is correct," or something similar. There are only two explanations. Either he did not discover that casting out nines does not infallibly prove a calculation is correct during his more than twenty-fours years of study, which does not speak well for his mental powers, or he knew it and suppressed it in his book, which does not speak well for his ethics.

S. held a press conference to publicize his book, reporters came, and at least one story appeared in a newspaper. Two readers then wrote letters to the editor to inform the newspaper, its readers, and S. that casting out nines was well known, and appeared, among other places, in Trachtenberg's *Speed System of Basic Mathematics*. S. replied:

> I was never aware of the existence of Professor Trachtenberg's book during the 25 years of my research into Instant Mathematics. My research was based entirely on my own findings and evaluations. I should get hold of this book someday and read it to understand what it is all about.

S. defended himself further, and concluded by being modest and by wrapping himself in the flag, in this case Malaysia's:

> The promotion of Instant Mathematics is not for my personal pride or glory but for the country. This was why the programme at my first Press conference held on Jan 9 was set up to project the image and the glory of the country.
>
> Furthermore, being modest, I did not put my photograph on the cover of my book as many of my friends thought I should have. I would prefer to go around without being recognized as the author of *Instant Mathematics*.

The respectful and objective treatment given to cranks in newspapers—presenting both sides of a question when there is only one side—leads one to wonder how much other such material appears in their pages.

NONAGONS, REGULAR

Regular polygons are pretty, so naturally we want to be able to make them in a pretty way, namely with compass and straightedge alone. Equilateral triangles and squares are no trouble, but most of us would be stuck if we were asked to draw a regular pentagon. However, it is not hard once you know how: take a circle, draw a radius OC perpendicular to the diameter AB; join B to D, the midpoint of OC; bisect angle BDO to find E, and draw a perpendicular from E to the circle to get P. Then BP is one side of the pentagon, and stepping that distance off around the circle gives the finished polygon, as in Figure 29. To show that this construction is correct is also not hard. If the circle has radius 1 and center at the origin, then the coordinates of D are $(0, 1/2)$, and $\tan 2\theta = 2$. From the perhaps not so well-known formula

$$\tan 2\theta = \frac{2 \tan \theta}{1 - \tan^2 \theta},$$

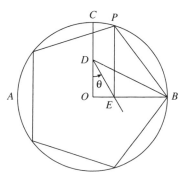

FIGURE 29

it follows that

$$2(1 - \tan^2 \theta) = 2 \tan \theta.$$

Solving that gives

$$\tan \theta = \frac{\sqrt{5} - 1}{2},$$

so

$$x = \frac{\tan \theta}{2} = \frac{\sqrt{5} - 1}{4},$$

and that is $\cos 72°$.

Stepping around the circumference of a circle with its radius divides the circumference into six equal parts, so we can easily get a regular hexagon. An octagon can be constructed by circumscribing a circle around a square and then drawing some perpendicular bisectors. Thus, we can make regular polygons with 3, 4, 5, 6, and 8 sides. What about those with 7 or 9? Those cannot be made with straightedge and compass. An extra tool, such as a crank, must be used. The problem of constructing a regular septagon with straightedge and compass alone has for some reason not attracted the attention of cranks, except those who think they have found methods for dividing any angle into any number of equal parts, seven included. There are not many nonagoners either, because a nonagoner would have to choose not to be a trisector. Nonagons follow as corollaries from trisections, being easily made by trisecting $120°$ angles, as in Figure 30; and it would an odd crank indeed who would pass up a famous and general problem for an obscure and particular one.

Nevertheless, nonagoners exist, and Figure 31 is a nonagon construction that was made independently of any trisection. On the circle, mark off $|AB|$ and $|AC|$ equal to $|OC|$, and draw arcs from O to A with centers at B and C, both with radius $|OA|$. Trisect OA at D, draw EF perpendicular to OA, and you have the side of the nonagon inscribed in the circle with radius $|OE|$. That is, the angle EOF is supposed to be $40°$, but it falls short by quite a bit since it measures only $39.6°$.

With nonagons as with so much else, you get what you pay for. Simple constructions give inaccurate results, and to get a decent nonagon you have to put some effort into the construction. Figure 32 is what I am sure is the best nonagon construction ever: draw arcs with radius $|AB|$ with centers at A and B to get C, and drop a perpendicular to locate O. An arc with center C and radius $|CA|$ intersects the perpendicular at D. Then, an arc with center A and radius $|DA|$ determines

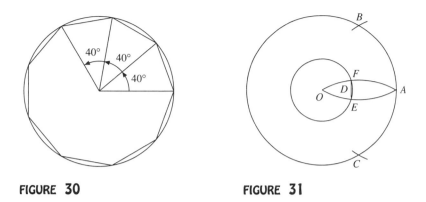

FIGURE 30   FIGURE 31

E, and a dropped perpendicular gives F and G. An arc with center G and radius $|GF|$ determines H, and a perpendicular dropped to AB locates I, very near to A. Next, an arc with center C and radius $|CO|$ determines J, and IJ intersects the perpendicular erected from B at K. Finally BFK is (almost) a $10°$ angle, so that by adding to it an easily-constructed $30°$ angle, we get the $40°$ angle that we need to make a regular nonagon. The construction, how discovered I have no idea, gives terrific accuracy: the angle departs from $10°$ by about .0001 seconds. So, if you were using it to inscribe a nonagon in the earth's orbit (assuming the orbit to be circular, with radius 93,000,000 miles), its side would be in error by around 240 feet. No one would notice.

When nonagoners explain why their constructions are correct, they tend to rattle on much as trisectors do:

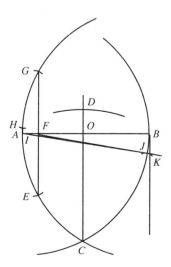

FIGURE 32

The history of this classic constructibility problem spans more than 2000 years of frustration and nearly 200 years of unwarranted assumption. That assumption is based on an error—not of mathematics but of logic— that was inadvertently made by Gauss himself in the concluding sentence of his Article 365. There, he issued a warning against even *attempting* to achieve a large class of geometric constructions, including that of a regular nine-sided polygon: the regular nonagon.

Gauss's error can be discovered in his unstated—but logically required—premise necessary for the validity of his warning itself. Reconstructed, the premise is:

Any solution that cannot be precisely examined and rigorously validated by traditional analytic geometry (founded on traditional trigonometry) *cannot* be precisely determined and rigorously validated by a nontraditional analytic geometry (founded on any nontraditional trigonometry —presently known or unknown).

The first seven pages of this paper have presented and validated what is in fact a nontraditional (because unique and unprecedented) analytic geometry. It is now evident that analytic geometry is a genus with at least two species. One species is *coordinate* ("Cartesian") analytic geometry. The second species I have titled indicative analytic geometry. . . .

and off we go.

That nonagoner was a crank of the third kind. Cranks of the first kind say "I am right." Period. No reason necessary, no reason given. Cranks of the second kind say "I am right, because . . . ," and proceed to try to justify their work with mathematics. Cranks of the third kind recognize that what they have done conflicts with mathematics and therefore, since either they or mathematics are incorrect, they try to change mathematics to fit what they have accomplished. There are no cranks of the fourth kind.

NOTATION, NONSTANDARD

Our mathematical notation, arrived at after centuries of evolution, is something that users of it unconsciously assume is as close to perfection as makes no difference. Clearly,

$$x^3 - 3x^2 + 17 = \sqrt{x^2 - 36}$$

is as superior to

$$R.3^a \tilde{m}.3.ce.\overline{p}.\underline{17} \qquad Rce.\tilde{m}.36$$

(as Pacioli would have written it in 1500) as a Boeing 777 is to the *Nina*, the *Pinta*, or the *Santa Maria*. Any other way of writing the equation would be absurd.

Well, perhaps. Pacioli did well enough with his notation, and he could probably read and write it almost as quickly as we do ours. Although mathematical notation has remained essentially unchanged for more than three hundred years, it is possible that future mathematicians will look back with amusement and pity at our notation: amusement because the symbols look so *funny*, and pity because of the time we wasted struggling with clumsy notation and because of the things that we did not see that are obvious once the proper notation is used. We are prisoners of our time and of our notation, but we do not notice our chains.

Mathematical amateurs and cranks are not as fettered as those who have become part of the mathematical community and have made their own all of the assumptions that the community makes. Amateurs sometimes use new notation. The natural reaction to new notation introduced by an amateur is that its inventor is not part of the community, and whatever is being introduced is not worth looking at. The amateur thus does not get any encouragement, and the crank gets confirmation of his cranky thoughts about conspiracies to keep down the work of outsiders.

For example, consider the following, written by W. B. in 1978:

> My works include several extended progression relationships. For example, the anti-derivative of the product of two functions:
>
> $$(fg)^* = c + f^*g - f^{**}g' + f^{***}g'' - \cdots$$
>
> when $g^{(n)}$ approaches zero. (The Taylor series is a special case.)
>
> I have also produced some root-approximation formulas:
>
> $$A^{1/e} \approx \frac{(e-1)c^e + Ad^e}{ec^{e-1}d}$$
>
> $$\overline{z\sqrt{}} \approx \frac{\ln z + e}{\ln e + 1}.$$

B. then went on to mention his discoveries in higher algebra, not stopping to explain his formulas. The first one gives no difficulty: it is integration by parts applied repeatedly. The idea is not new, and it even appeared not too long ago in *The College Mathematics Journal* as a pedagogical aid. The second formula would cause anyone to pause and ask, "Whatever are c and d?" The third would cause anyone to give up: even if e isn't $2.7182818\ldots$, that backwards square-root sign has to be defined.

B. may have been at fault for not defining his symbols, but his three formulas are not nutty. The second one gives a method for going from one approximation to the eth root of A, c/d, to a better one. Expressed in the "proper" mathematical form, B. said that with appropriate starting values a_0 and b_0, the sequence converges to $\sqrt[n]{A}$, where

$$a_{k+1} = (n-1)a_k^n + Ab_k^n \qquad \text{and} \qquad b_{k+1} = na_k^{n-1}b_k.$$

When $n = 2$, the sequence is

$$\frac{a_{k+1}}{b_{k+1}} = \frac{a_k^2 + Ab_k^2}{2a_kb_k}.$$

If we take $A = 3$ with first approximation $2/1$, the formula gives

$$\frac{7}{4}, \frac{97}{56}, \frac{18817}{10864}, \ldots$$

as approximations to $\sqrt{3}$. They are not bad: squaring the last one gives 3.0000000047.... The reason they are not bad is that what B. was generating were continued fraction approximations to $\sqrt{A}$, since

$$a_{k+1} + \sqrt{A}b_{k+1} = (a_k + \sqrt{A}b_k)^2.$$

That explains why his sequence of square-root approximations converges. But why do his cube-root sequences

$$\frac{a_{k+1}}{b_{k+1}} = \frac{2a_k^3 + Ab_k^3}{3a_k^2 b_k}$$

converge? It cannot be because of continued fractions, since cube roots do not have nice continued-fraction expansions. Did B. find something new?

Everything is clear in his third formula after B. explained, finally, that z enclosed with a backward square-root sign means the solution of

$$y^y = z,$$

and e stands for "estimate." So we have another sequence of better and better estimates,

$$a_{k+1} = \frac{\ln z + a_k}{\ln a_k + 1}.$$

This formula also works pretty well. If you are curious to know for what value of y the equation $y^y = 10$ is satisfied—something that your calculator will not tell you—start with a reasonable guess like 2.5 ($2.5^{2.5} = 9.88\ldots$) and B.'s formula will give you 2.5061881..., which raised to its own power is 10.000076.... The reason the formula works is that the equation to be solved is $y \ln y = \ln z$. That can be written as $y(\ln y + 1) = \ln z + y$ or as

$$y = \frac{\ln z + y}{\ln y + 1},$$

which gives the recursion relation.

How should B. have been treated? I think that he should have been given a pat on the back for being clever. He was clever, and complimenting him for it does no harm and is good public relations. He would feel good about mathematics and mathematicians, and his attitude might rub off on those close to him. Mathematicians are sometimes given to bewailing how unappreciated they are by the general public, and part of the reason for that, I think, is the lack of appreciation that they have for the general public.

This is what B. had to say about how he was treated:

> For many years I have been working on my own in the field of mathematics. During this time, I have developed some new formulas and teaching techniques which could greatly benefit the educational system. Unfortunately, my attempts to inform teachers or college professors of these discoveries have been unsuccessful since they do not have the time or are unwilling to listen.

Later he wrote:

> What my years of experience with mathematicians has taught me is their lack of objectivity.

He concluded:

> Now I ask you not to be resentful or unhappy over what I am about to write. My feelings are that I would be ashamed to call myself a mathematician.

A potential mathematics fan, a possible booster for the subject, turned sour with ill feelings toward the mathematics community! It did not have to be. The mathematics community ought to consider following the example of those companies for whom the customer is always right. Even when the customer relations department of such a company knows in its heart that the complaining customer is at best unjustified and at worst a cheat, it grits its teeth, refunds the money, and sends a mollifying letter enclosing a free sample. The good will, it thinks—the customer will tell everyone about how nice the company is—makes it worthwhile. Mathematicians, though, are so unswervingly devoted to truth that it is possible that no amount of tooth gritting could bring them to tell B. anything more than that he is an amateur who has rediscovered a few things of no account.

NUMBER THEORY, THE LURE OF

What would you think if you saw a magazine cover containing the following head-lines?

How the government could reduce pollution with the stroke of a pen.

Is somebody out to get you? A new defense.

What do the Chinese really think they're doing?

More light on Black unrest.

Nothing much, probably, beyond that it sounds like just another general-interest, current-events magazine trying to sell copies. However, those headlines appeared on the front cover of *Hypermodern*, a periodical written, edited, and published by one person, A. C.

Usually anyone who puts out his own magazine qualifies for the label "crank," and C. was no exception. For example, this is what he had to say about Stonehenge:

Now, suppose you take a board and drive 30 nails into it, roughly in a circle, roughly evenly spaced. What do you have now? You have 435 peepsights, since 30 points form 435 distinct pairs of points. Now you may fix your board level somewhere where you have a view of the horizon, and you have about one peepsight for every four-fifths of a degree, on the average.

Those peepsights should be enough to line up fairly closely on any phenomenon that takes place on the horizon, even though the nails may not have been placed carefully and you may not have had any astronomical use in mind at all.

Actually, since there are 30 stones in the circle at Stonehenge, there are 60 reference points, 30 centers of the stones and 30 centers of the points between them. Since, however, as a practical matter, most sightings from a stone to another stone nearby on the arc would not be very good, because of the thickness of the stones, let us suppose that we sight only from any point to the one-half of the points on the semi-circle directly opposite the chosen point. Working in this way around the whole circle, we get 60×30 peepsights, or 1800, roughly five for every degree of arc.

Thus, the prehistoric pillars of Stonehenge may form an astronomical observatory by sheer overkill, without planning, care, or even intention.

In these days it is cranky *not* to attribute deep astronomical knowledge to the primitive and probably savage builders of Stonehenge.

The lead story in issue number 12 of *Hypermodern* is not about Stonehenge, pollution, the Chinese, or defending yourself against people out to get you; it is about number theory. Mathematics is a magnet attracting minds, and number theory is its most powerful attractor. C. presented forty-five of his number-theoretical conjectures. Here is a sample:

2. Numbers divisible by 7 are never equal to a square plus twice a square, unless they are divisible by 49.

11. Every prime of the form $20x + 1$ or $20x + 9$ is the sum of a square and five times a square, and in one way only.

15. Every number of the form $4x + 3$ is the sum of an odd square and twice a prime, except 79.

22. Squares not divisible by three may be written as the sum of a square and a triangular number.

23. The only cube which equals a square plus a triangular number is 64.

27. Every even number is the sum of a Fibonacci number and a prime, except 148 and 208.

33. Every number divisible by four is the sum of a prime of the form $4x + 1$ and $4x + 3$.

41. Every odd number is the sum of a prime and four times a prime, except 77.

The conjectures range from things that are known, such as number 11, to things that are unknown and likely to stay that way for quite some time, as the Goldbach-like conjecture of number 33. Number 22 is easy, an exercise in finding an algebraic identity; number 23 is hard and may be true; and number 27 is hard and probably false. C. presented them all indiscriminately:

We have some familiarity with the field, but do not have time to chase down the appropriate texts, much less read them (this is an originating publication, not a research publication).

The power of number theory! Amateurs do not make conjectures in topology, or in partial differential equations.

I would bet that conjecture number 15 is false, and it would be fun to find a counterexample. However, since I think that *Hypermodern* ceased publication some years ago, it would be impossible to correct for its readers any erroneous impression that the conjecture left with them.

PERFECT NUMBERS

The ancient Pythagoreans said that all was number. A corollary to that is that the more you know about numbers, the more you know about the All. Hence it is a good and worthy occupation to contemplate numbers. Some of the Pythagoreans' contemplation was devoted to how numbers could be divided into parts. The prime numbers were those that could not be divided into parts. For those numbers that had parts, it is not surprising that, over years of contemplation, some Pythagorean had the idea of adding the parts together. Most small integers have a sum of parts that is smaller than the whole: $1 + 2 + 4 < 8$, $1 + 2 + 5 < 10$. Some have a sum that is larger: $1 + 2 + 3 + 4 + 6 > 12$, $1 + 2 + 3 + 6 + 9 > 18$. And one small integer is equal to the sum of its parts: $1 + 2 + 3 = 6$. The Pythagoreans classified an integer as deficient, abundant, or perfect depending on whether it was less than, greater than, or equal to the sum of its parts. They then proceeded to give mystical significance to the terms, to help them in their understanding of the All that was number.

Mathematicians no longer care about the mystical significance of numbers, but deficient, abundant, and perfect numbers raise interesting mathematical problems. 12, 18, 24, and 30 are abundant; are $36, 42, 48, \ldots$ all abundant? (Yes, they are.) Is every abundant number a multiple of 6? (No, 40 is not.) Well, is every abundant number even? (No, 945 is abundant.) What are the perfect numbers after 6? (The next three are 28, 496, and 8128.) The next perfect number after 8128 is too big to be found by non-systematic searching, so something else is needed, something that might be found by looking at 6, 28, 496, and 8128 and seeing what they have in common:

$$6 = 2 \cdot 3, \qquad 28 = 2^2 \cdot 7, \qquad 496 = 2^4 \cdot 31, \qquad 8128 = 2^6 \cdot 127.$$

What they have in common is that each is a power of 2 times a prime. The ancient Greeks noticed that the numbers are

$$2^{2-1}(2^2 - 1), \qquad 2^{3-1}(2^3 - 1), \qquad 2^{5-1}(2^5 - 1), \qquad 2^{7-1}(2^7 - 1).$$

They then guessed that if p and $2^p - 1$ are prime, then $2^{p-1}(2^p - 1)$ is perfect. They also proved it, and a proof is included in Euclid's *Elements*.

Two thousand years later, Euler proved that those numbers are the only even perfect numbers. No odd perfect numbers are known, and the general opinion is that there are none. Conditions have been found that odd perfect numbers must satisfy, conditions that imply that the smallest odd perfect number, if there is one, must be very, very large—a number with at least 150 digits, among other properties. However, a proof that there are no odd perfect numbers is not in sight.

The search for perfect numbers does not attract the same attention outside the mathematical community that trisecting the angle or squaring the circle does, but word about perfect numbers has gotten out. L. S., a mathematical amateur who retraced the steps of the ancient Greeks, found (he thought) that $2^{p-1}(2^p - 1)$ with p prime is always a perfect number, and he bothered a scientific magazine with his discovery. Since he was an amateur and not a crank, the editor's reply was more or less appropriate:

> I suggest you write to the head of the mathematics department [of a nearby college], explain your need and ask whether he would mind having one of the younger members of the faculty who may not be too busy look over what you have done and advise you as to its merits.

An assistant professor at the college politely pointed out that $2^{10}(2^{11} - 1)$ is not perfect because $2^{11} - 1 = 2047 = 23 \cdot 89$ is not prime. He gave S. some references to books on number theory (something that assistant professors often do, though the references are almost always over the heads of the amateurs), and congratulated him on his interest in mathematics. If that had been the end of it, S. would not belong in this book, but it was not: S. changed from amateur to crank. He replied with more than one letter asserting, for reasons that were impossible for me to understand, that his number *was* perfect and his formula worked every time. That was illogical, as was his proof that there are no odd perfect numbers:

> The form of the Equation for generating Perfect Numbers implies that *all* Perfect Numbers are *Even*, since 2 to any Power is Even. Hence no odd Perfect Numbers are possible.

That is, he said in effect that since *his* formula gives even perfect numbers, there are no odd ones. Proceeding similarly, I could conclude from *my* formula for abundant numbers, $A = 6p$ for a prime $p > 3$—abundant since the sum of the parts of A,

$$1 + 2 + 3 + 6 + p + 2p + 3p = 6p + 12,$$

is greater than A—that there were no odd abundant numbers. However, the conclusion does not follow, as the existence of 945 shows.

I do not know if S. proceeded further down the crank road, going on to pester other mathematicians. His last letter ended:

> I am sorry I have to make this long-hand, but I have no typewriter and have not the means to get one.

Let us hope that his hand got tired.

This number,

$$\phi = \frac{1 + \sqrt{5}}{2} = 1.618034\dots,$$

has several properties. For one thing, if you enter it into your pocket calculator and push the reciprocal button, none of the digits will change, except that the first "1" will be replaced with "0." That is,

$$\frac{1}{\phi} = \phi - 1.$$

Another way of writing that is

$$\phi = 1 + \frac{1}{\phi},$$

from which we get

$$\phi = 1 + \frac{1}{\phi} = 1 + \frac{1}{\left(1 + \dfrac{1}{\phi}\right)},$$

and that is equal to

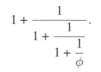

Continuing,

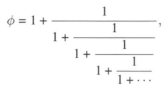

which is nothing if not picturesque.

Something that simple must somehow be built into nature, and ϕ is. If you truncate the continued fraction for ϕ you get

$$1 + \frac{1}{1} = \frac{2}{1}, \qquad 1 + \frac{1}{1 + \frac{1}{1}} = 1 + \frac{1}{2} = \frac{3}{2},$$

and the next fractions will be $\frac{5}{3}, \frac{8}{5}, \frac{13}{8}, \ldots$: ratios of successive Fibonacci numbers. Fibonacci numbers describe, among other things, how sunflower seeds arrange themselves in spirals. Also, $\phi = 2\cos 72° + 1$, so ϕ is just what you need to draw that mystic figure, the pentagon (Figure 33). Further, the most aesthetically pleasing of all rectangles, shown in Figure 34, has sides whose ratio is ϕ. Phi is everywhere.

However, it is not quite as widespread as R. G., an extremely prolific writer, would have it. He claimed that

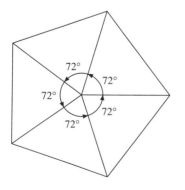

FIGURE 33

FIGURE 34

This UNIVERSAL PARAMETER manifests in ALL natural phenomena INCLUDING CONSCIOUSNESS!!!

As evidence for that, he cited what he said Plato said in the *Timaeus*:

Phi is the key to the Physics of the Universe.

Quite a few of the equations of physics involve π—G. mentioned those for the period of a pendulum, Planck's constant, and a strange one for "energy,"

$$\pi = \left(\frac{E}{1/2\,mc^2}\right)^{1/2}\left[J_\lambda \cdot \lambda^5(e^{hc/k\lambda T} - 1)\right].$$

Since ϕ is fundamental, G. turned them into equations involving ϕ instead of π by using the relation he had found between them, namely

$$\frac{6}{5}\phi^2 = \pi.$$

This is not a bad approximation: it says that

$$\pi = \frac{3(3 + \sqrt{5})}{5} = 3.1416408 \ldots.$$

G. offered no justification for his value, nor did he comment on its difference from the traditional $3.14159\ldots$. Besides π, physics formulas involve e, but G. was prepared. The es can also be replaced with ϕs, since, G. said,

$$e = \phi^2 + \frac{1}{10} \quad (= 2.718034\ldots).$$

G.'s writings are all bound up with the great pyramid in Egypt and 52.36, the "royal cubit" used by pyramidologists in getting some of the multitude of numbers that they manipulate to make their discoveries. G. wrote:

I have an engineering and architectural background. I began this study with the conviction that the ancient cultures were extraordinary and that the GIZA PYRAMIDS especially were supernally unique artifacts. My preoccupation with the physical aspects of this study were continually directed towards the principles inherent in the construction of these PYRAMIDS and other world-wide SACRED STRUCTURES. ALL OF THE RESEARCH PATHS CONTINUALLY SPIRALLED TOWARDS THE SACRED SIGNIFICANCE OF THE PHI PARAMETER.

Since $20\phi^2 = 52.36$, the pyramid builders were quite aware, G. said, of the importance of ϕ. They were also aware of the distances of the planets from the sun, even the ones that had not been discovered when the pyramids were under construction. G. had a table, with distance measured in astronomical units:

Planet	Distance from Sun	Reason
Mercury	.387	2/5236 = 382
Venus	.723	4/5236 = 764
Earth	1.0	5236/5236 = 1
Mars	1.52	8/5236 = 1528
Jupiter	5.2	27/5236 = 5157
Saturn	9.47	50/5236 = 9550
Uranus	20.	1618/5236 = 19864
Neptune	31.	104/5236 = 30901
Pluto	39.59	208/5236 = 39728

G. ignored decimal points in favor of significant digits, and the numerator for Uranus comes from the first four digits of ϕ.

G. kept up with scientific developments, so when he read an account of fractals in which $\ln 4/\ln 3 = 1.2618\ldots$ appeared as the fractional dimension of something, he was quick to note that

$$\frac{\ln 4}{\ln 3} = 1 + \frac{\phi^2}{10}.$$

All this is harmless enough, but G. also spent time at the typewriter, generating letters to people who did not necessarily need letters from him. He wrote to *Scientific American* columnists Martin Gardner and A. K. Dewdney. Benoit Mandelbrot, whose book on fractals mentioned the constant .7 in an example, got a letter from G. asking if the constant ought to be $\phi - 1 = .618\ldots$ instead. Sometimes G. got answers: the Foundation for the Study of Cycles wrote:

> Thank you for sending us your material on the Golden Mean.
> It makes a valuable addition to our library.

A person in the Elliott Wave organization answered, amazingly:

> I apologize for the late reply, but am just getting around to answering a large stack of mail. I did appreciate your thoughts from the letter you wrote last fall. Unquestionably, the universe is tied together according to a few simple principles. *Phi* is one, if not the main.

That did in the Elliott Wave Theory for me. I have decided, in spite of knowing almost nothing about it, that there is a good chance that it is crankery.

It is also amazing that there were people who took G.'s ideas seriously. One of his correspondents forwarded his material to a third party, who replied:

> Regarding [G.]'s letter, I am somewhat familiar with his approach, being that of the classic theosophical view of dynamic symmetry and "fourth-dimensional geometry." There are connections with synergetics. Here very briefly are some of them.
>
> The theosophical attempt to break out of limitations of two-dimensional modeling of their theories led to a renewed interest in dynamic symmetry in the 1920s and in some of its more perceptive exponents to an omnidirectional nested model (unfettered by the limitations of reexamining the surface pentagrams). Using a somewhat unsynergetic, backward approach (I give you tetrahedron with edge-length 1.414), the proportions of the Golden Section exhibit themselves most elegantly as:

tangently nested	edge-lengths
Inner Icosahedron	.382
Dodecahedron	.618
Outer Icosahedron	1.618
with Cube being	1.000

> This model's asymmetry, I would guess, produces a dual spiral. The contemporary theosophist, L. Gordon Plummer, describes such a model as involution and evolution within the tetraktys. I find this model, as incomplete (primitive?) as it is, in some ways even more revolutionary than Fuller's symmetrical contractions of the icosahedron in jitterbugging and right-left phase shifts of the icosahedron through the vector equilibrium.
>
> If I were to explore this idea further, I would look for an inherent fiveness (*phi*veness?) in the concept of domain. This elusive fiveness may hold the key. Especially when considered in terms of contained volume rather than surface areas. What the theosophists inadequately described (but were unable to model) as hyperspace (hyperplane, hypervolume, etc.) and, perhaps, what Fuller models as the omnidirectional domain of points, may relate to biological processes (Fuller says "biospherical patterns," but also includes the nonbiological). The modular subdivisions of spherics to higher symmetry and a complementary but asymmetric process may relate to both eternal regeneration and the dynamic process of ephemeralization (Dymaxion). Further, it may relate to what students of Gurdjieff called "the reverberating chord"—the syntropic process occurring on the

surface of the Earth (the half-note between *mi* and *fa* on the musical scale) (transducer of *fa* energy to *mi*), as well as the second (?) aspect of the tri- partite *do* of the musical scale (the Gurdjieffians say "ray of creation").

Another line to explore would be sphere to space to sphere transfor- mations. Perhaps an oversimplification but beautiful nevertheless: sphere (radiation-matter); space (gravitation); isotropic vector matrix (Unified Field).

The letter goes on further. I did not know that G.'s writings fit into an evidently large structure, engaging the attention of many people of intelligence (no stupid person could have composed the preceding letter) who were busy searching out new and deeper results. It is just like mathematics! Or perhaps there are differences.

PRAYER, MATRIX

J. D., a priest of the Church of England, wrote in 1977:

> I have long been accustomed to think of God and pray to him in mathematical terms, and one (but only one) such method involves the use of matrices.
>
> Now that children and young people in my congregation are learning matrices from the age of 13, I am starting to teach them how to pray and think of God in matrices. As one young lady of the congregation said to me the day before yesterday "There's nothing difficult about it".
>
> Unfortunately the senior clergy and established theologians all seem to be unmathematical. I have been quite unsuccessful in trying to get the editors of "Theology" interested in mathematical theology. I have been trying to teach my Bishop how to pray in matrices, the only result so far is that after a recent Confirmation, when the Bishop was talking to relatives of the candidates, a grandmother said "It's all right, Bishop, we understand Mr. [D.]", and he replied "That's more than I do"!

D. gave examples. The first is a matrix prayer for peace in Northern Ireland.

> According to St Paul "The peace of God is beyond our utmost understanding" and is offered to us. Quantify it therefore in terms of some field F known to God but unknown to us and represent it by the column vector $\{S\}$ of infinite dimension (S for Serenity, we cannot use P, it is needed for a special purpose). The elements of $\{S\}$ are finite because we are finite.
>
> St Paul tells us that it will "keep guard over your hearts and your thoughts in Christ Jesus", the original Greek for "your" is plural. So

represent the social action of the peace of God on a group of people "in Christ Jesus" by premultiplying $\{S\}$ with a matrix that has as many rows as there are people in the group and an infinite number of columns, the "revelation matrix", R. The product we will call $\{U\}$, the peacemaking power of the Christian people of all denominations in Ulster. So

$$R\{S\} \to \{U\}$$

If there are 1,500,000 Christians of all denominations in Ulster R will have as many rows. The row m for some particular Mr Murphy will have some elements zero, or with special positive or negative finite values, corresponding to matters that particularly concern him, the other elements will be members of some absolutely convergent series, arranged so that each element has a value corresponding to Mr Murphy's ability to receive God's revelation of peace for some aspect, opportunity or action of Mr Murphy's "heart and thought". His free will can affect a finite number of the elements of m but cannot overrule God's requirement that the infinite series of elements in m shall be absolutely convergent. Our free will is really free though limited, God's power is infinite.

$\{U\}$ will have as many elements as there are Christians in Northern Ireland, one element, u_m, will represent the peace in heart and mind of Mr Murphy. Suppose he feels aggressive and unpeaceful whenever he thinks of the Battle of the Boyne, then r_{mb} is negative, corresponding to s_b, the element of S representing that aspect of the peace of God which can keep guard over the hearts and thoughts of people thinking about the Battle of the Boyne. We should pray that r_{mb} may become positive. Any change of sign from negative to positive in an element of m will increase the value of u_m, Mr Murphy's whole peace of heart and mind; and therefore will also increase the total peacemaking power of the Christians of Ulster.

$R\{S\} \to \{U\}$ is therefore an accurate and completely detailed statement of how the peace of God is keeping guard over the hearts and thoughts of all the people in Christ Jesus in Northern Ireland. To transform it into an intercession we need a Polite Request Operator, I suggest

$$P\langle \dots \rangle$$

where the operator acts on everything within the pointed brackets, and is read as "Please cause". So here is the prayer:

$$P\langle R\{S\} \to \{U\}, \text{ all } r > 0\rangle.$$

This prayer should be sufficiently concise to be acceptable to Christ, yet every single Christian inhabitant of Northern Ireland has been separately included with detailed consideration of all the infinite ways in which the peace of God can, if he lets it, keep guard over his heart and thought.

I think that is a perfectly splendid application of mathematics. It gives a mathematical model out of which only good can come. However, it shares a difficulty with many other mathematical models, namely that it is very hard or impossible to determine the numerical values of all the quantities in it. There are models in mathematical economics that, if we only knew precisely what peoples' utility functions were, or their subjective probabilities, then we would be able to make accurate predictions. But we do not (or cannot) know them precisely, so the predictions of mathematical economics continue to be variable and unreliable. Similarly, the numerical values in D.'s matrices would be hard to specify. What goes into the rows of S? The units in which to measure serenity have not been determined.

Another difficulty, just as large in a different way, is that linear algebra is not often part of the training of future members of the clergy. In fact, if you threw a rock at a random congregation you might be more likely to hit someone who knew how to multiply two matrices together than if you threw it at an equally numerous gathering of the clergy. So, it is highly unlikely that matrix prayer will ever become popular, or even survive its originator. Too bad: it is sufficiently novel that God might pay more attention to it than to the more usual supplications.

The field of mathematical theology is potentially rich. Introducing another of his examples, D. wrote:

> Represent heaven, the home of God, as a vector space of infinite dimension over some field F known to God but unknown to us, in which the activities of God are quantifiable. Lengths will be measured (Revelation chapter 21 verse 15) in the usual way, as the square roots of the inner self-products of vectors (assuming heaven to be euclidean).

The geometry of heaven! Does it follow the rules of Euclid? What if it doesn't? There is much room for fascinating speculation.

PRIMES, THE SECRET OF THE

Many people have sought the secret of the primes, but the primes are not about to give it up. The primes are genuinely mysterious, so mysterious that in the literature of crank mathematics they do not appear very often because it is not easy to say anything about them that is both cranky and not obviously false.

Nevertheless, one attempt is the *Prototype Table of Prime Numbers*, by N. Y., written in 1981. Y. noted that primes, once you get past 2 and 3, leave remainders of one or five when divided by 6. She called the two classes "heptals" or "pentals," made some observations, such as the product of two heptals is a heptal, and constructed a table:

$6n - 1$	5	11	17	23	29	c	41	47	53	59	c	...
n	1	2	3	4	5	6	7	8	9	10	11	...
$6n + 1$	7	13	19	c	31	37	43	c	c	61	67	...

("c" stands for composite.) She concluded:

> This $6n$ progression is really the axle that rules the twin function-ing of the pental and heptal progressions. Since in the terms of the twin progressions are found all the primes greater than *two* and *three*, and also all the composites that are not produced by *two* and *three*, it leads to the premise that the n's are the control in a *uniform rule* for factorization of the pental and heptal terms.

But she did not give the uniform rule.

Y.'s work is more a tribute to the attractive power of mathematics than it is crankery. Y., who put "PhD" after her name, was an Associate Professor of Sociology at a western state university. Mathematics is more fascinating than sociology: as far as I know, no mathematician has ever claimed to have found solutions to important unanswered questions in sociology. Mathematicians do not even become engrossed in the field as a hobby. The traffic is all the other way, just as some college students change their majors from mathematics to sociology, but none does the reverse.

PRIMES, TWIN, EXISTENCE OF INFINITELY MANY

As mentioned previously, prime numbers are fascinating. They occur among the integers seemingly at random, though they are not random at all. They thin out as you progress through the integers, and the farther we go toward infinity, the fewer we find. But, as Euclid proved, they never disappear. Nor, despite their thinning, do twin primes—two consecutive odd numbers, both primes—disappear:

$$3, 5; \quad 5, 7; \quad 11, 13; \quad 17, 19; \ldots 1159142985 \cdot 2^{2304} \pm 1; \ldots;$$

they keep coming, however far out we go among the integers. Everyone believes that there are infinitely many twin primes, but no one has yet been able to prove it.

In 1919, Viggo Brun had the idea of looking at the sum of the reciprocals of the twin primes. If he could have shown that the sum diverges, then we would know that there are infinitely many twin primes, since a finite series cannot diverge. However, what he proved is that the sum of the reciprocals of the twin primes converges. This is consistent with there being only finitely many twin primes, but it is also consistent with the existence of infinitely many. His proof is not even very strong evidence against the existence of infinitely many twin primes. The sum of the reciprocals of the primes diverges, but so slowly,

$$\sum_{p \leq n, \; p \text{ prime}} \frac{1}{p} \approx \log \log n,$$

that it is no surprise that when you eliminate most of the primes (most primes are not twin primes),

$$\sum_{p,\,p+2 \text{ prime}} \frac{1}{p}$$

converges. The series sums to $1.90216054\ldots$, a number that may be rational or irrational, algebraic or transcendental, and also has yet to be investigated for its numerological content.

Brun's proof is neither trivial nor easy, and one reason for its difficulty is the mixture of addition and multiplication in the theorem. The primes are the building blocks out of which all the integers may be constructed by multiplication, but we go from p to $p + 2$ by addition. There is no obvious reason why multiplicative objects should have any additive structure, so the proof of the existence of infinitely many twin primes lies deep, deeper than anyone has yet gone.

The difficulty of the proof does not keep people from trying to find it, which is good; but sometimes they think that they have succeeded when they have not, which is not good. Even professional mathematicians can be fooled, as was Professor C. M. I have a photocopy of a newspaper clipping that reads, in its entirety:

EUCLID'S PROBLEM SOLVED AT LAST

WELLESLEY, Mass.—Euclid, the great ancient Greek mathematician who lived 2200 years ago, presented a classical problem that he was unable to solve and which remained unsolved until the present time. A solution was presented here recently at the meeting of the American Mathematical Society by Dr. [C. M], professor of mathematics, University of [C.]. The total number of unsolved mathematical problems is now decreased by one.

Euclid, Dr. [M.] stated, proved that there exist an infinite number of primes, that is, numbers such as two, three, five and seven having no divisors but themselves and unity. Succeeding generations of mathematicians have guessed, but have never been able to prove, that there likewise exist an infinite number of prime-pairs, that is, successive primes which differ by two, such as 11, 13; 17, 19; 41, 43, and so on.

At the meeting Dr. [M.] presented an involved but convincing paper giving his proof of the infinitude of prime-pairs.

Pretty good, as newspaper reporting of mathematics goes, though it was too bad that the headline writer was not as accurate as the reporter.

The story is undated, but the photocopy includes the end of the story above Dr. M.'s:

... normal driving needs, but is not sufficient to withstand a rush to hoard gasoline not needed now.

Some gasoline originally intended for this district has been sent East for immediate shipment overseas, but there is enough to meet actual needs, it was stated.

It also includes the beginning of the story below:

GET FULL VISION

WASHINGTON—Observation panels of bulletproof glass installed in turrets of Allied tanks are known. . . .

Thus, the clipping dates from World War II and was taken from a midwestern newspaper. Can anyone recall it? Can anyone remember being in Wellesley 45 years ago and seeing the twin primes theorem being proved by Professor M.?

One reason so few cranks bother with the twin primes problem is that there is no way of getting a handle on it. To square the circle, you sit down with compass and straightedge, draw a circle, and have at it: it is easy to be *doing* something. But there is not much that can be done with twin primes other than sit and look at them. In 1988, R. H. did the best he could when he said pick two numbers with no common factors, a and m, and make two sequences,

$$a + im \qquad \text{and} \qquad (a + 2) + im, \qquad i = 0, 1, 2, \dots.$$

By Dirichlet's theorem, he said, the first sequence contains infinitely many primes, which is true enough. Then, H. said, if $a + 2$ and m have no common factors either (you can always find a and m so that both a and $a + 2$ have no factors in common with m, such as $a = 5$ and $m = 12$), then the second sequence also has infinitely many prime members. Quite right also: for $a = 5$ and $m = 12$, the primes (italicized) come thick and fast:

$$5, \textit{17}, \textit{29}, \textit{41}, \textit{53}, 65, 77, \textit{89}, \textit{101}, \textit{113}, 125, \dots$$

$$7, \textit{19}, \textit{31}, \textit{43}, 55, \textit{67}, \textit{79}, 91, \textit{103}, 115, \textit{127}, \dots.$$

Then, H. said, look only at the prime members of the first sequence and the corresponding elements in the second sequence. Since there are infinitely many primes in the second sequence, by "the Self Similarity Axiom," infinitely many of them must appear in the subsequence, so there must be infinitely many twin primes.

The Self Similarity Axiom is:

The elements of an infinite subset of an infinite set contains all the recurring significant properties of that set unless the process that selects the elements of the subset directly excludes a property.

Unfortunately for H., that axiom is not recognized as one of the foundations of mathematics. How is it to be determined if a selection process directly excludes a property? Applied to the twin prime problem, the axiom says that there are infinitely many twin primes unless it happens that there are not. That is not a solution, nor is it progress.

PROLIFICITY, CRANK'S

Most cranks have but one string on their bows, which gets plucked over and over. The construction to trisect angles gets sent out, over and over; the proof of Fermat's Last Theorem, out, over and over; the squared circle, over and over. The details of the circle-squaring can change slightly, the fermatist's proof of FLT can (and usually does) become longer, and the trisector may add an appendix refuting Wantzel's proof that the trisection is impossible. But the crank's message remains the same. The reason, of course, is that it is rare for a person to have a new idea, even a crankish one; so when one comes it is natural to cling to it and work it for all it is worth.

There are very few cranks who have more than one subject or who produce more than a slim sheaf of work. A major exception is B. L., who has produced *thousands* of pages of work. He has produced them a page at a time, since that is his medium: the one-page, hand-written mathematical communication. Here is an example:

Summing e

$N!$	
1	1
3	2
10	6
41	24
206	120
1,237	720
8,660	5,040
69,281	40,320
623,530	362,880
6,235,301	3,628,800
68,588,312	39,916,800
823,059,745	479,001,600

The procedure, divide the denominator into the lower denominator, multiply the numerator, add 1, this is the new numerator.

$$823,059,745/479,001,600 = 1.718281828$$

My thanks to
　　my God   [B. L.]
                                                3/22/85

What L. was communicating on that sheet was that if you take his two columns of numbers, the right-hand column being values of $n!$ and the left-hand column values of u_n, where $u_1 = 1$ and

$$u_n = nu_{n-1} + 1, \qquad n = 2, 3, \ldots,$$

then the quotients $u_n/n!$ approach $e - 1$. Since

$$\frac{u_n}{n!} = \frac{u_{n-1}}{(n-1)!} + \frac{1}{n!},$$

it follows that

$$\frac{u_n}{n!} = \frac{1}{n!} + \frac{1}{(n-1)!} + \cdots + \frac{1}{2!} + 1,$$

a partial sum of the series for $e - 1$. L. thus put something well known into a not-so-well-known form. That is not crankish, but thinking that it is worthy of circulation is.

L. had spent some time looking into the literature of mathematics. Here is another of his sheets:

The Treasury of Mathematics
Edited by Henrietta O. Midonick
Philosophical Library, Inc. (1965)
Page 33

Archimedes Proposition 23

Given a series of areas $A, B, C, \ldots Z$ of which A is the greatest and each is equal to four times the next in order, then:

$$1 + \frac{1}{4} + \left(\frac{1}{4}\right)^2 + \cdots + \left(\frac{1}{4}\right)^{N-1} = \frac{1 - \left(\frac{1}{4}\right)^N}{1 - \frac{1}{4}}$$

Letter: 3/6/80.

$$\frac{1}{x^0} + \frac{1}{x^1} + \frac{1}{x^2} + \cdots + \frac{1}{x^{N-1}} = \frac{x^N - 1}{(x-1)(x^{N-1})}$$

$\therefore$ Archimedes

$$1 + \frac{1}{4} + \left(\frac{1}{4}\right)^2 + \cdots + \left(\frac{1}{4}\right)^{N-1} = \frac{1 - \left(\frac{1}{4}\right)^N}{1 - \frac{1}{4}}$$

Ltr: 3/6/80

$$x = 4, \quad 1 + \frac{1}{4} + \left(\frac{1}{4}\right)^2 + \cdots + \left(\frac{1}{4}\right)^{N-1} = \frac{4^N - 1}{(3)(4^{N-1})}$$

Now,

$$\frac{1 - \left(\frac{1}{4}\right)^N}{1 - \frac{1}{4}} = \frac{\frac{4^N - 1}{4^N}}{\frac{4 - 1}{4}} = \frac{4^N - 1}{4^N} \cdot \frac{4}{3} = \frac{4^N - 1}{(4^{N-1})(3)}$$

My thanks to
my God   [B. L.]
                                               4/28/80

That sheet shows that L.'s formula for the sum of a geometric series given in his letter of 3/6/80 agrees with the usual formula.

Besides the series for e and geometric series, L. had read about the technique of mathematical induction. He tried to use it to show that

$$\frac{1}{2} + \frac{2}{3} + \frac{3}{4} + \cdots + \frac{N}{N+1} \neq \frac{N^2}{N+1} :$$

Mathematical induction proves it is out of balance. A great "N" shows the power of the decimal.

$$\frac{1}{2} + \frac{2}{3} + \frac{3}{4} + \cdots + \overset{L_N}{\frac{N}{N+1}} = \overset{S_N}{\frac{N^2}{N+1}}$$

$$\text{Old } S_N + \text{New } L_N = \text{New } S_N$$

$$\frac{N^2}{N+1} + \frac{N+1}{N+2} = \frac{(N+1)^2}{N+2}$$

$$N^2(N+2) + (N+1) = (N+1)^3$$

$$N^3 + 2N^2 + N^2 + 2N + 1 = (N+1)^3$$

$$N^3 + 3N^2 + 2N + 1 \neq N^3 + 3N^2 + 3N + 1$$

Using "N" = 25000

$$\frac{625{,}000{,}000}{25{,}001} + \frac{25001}{25002} = \frac{625{,}050{,}001}{25{,}002}$$

$$24{,}999.00004 + 1^- = 25{,}000.0004$$

My thanks to
     my God

                    Seems I lied
                    [B. L.]
                    8/23/83

Why his proof by induction fails is a good exercise for beginning students of mathematical induction, as is the challenge of finding a direct proof.

L.'s pages, all with mathematics at the same level as the examples above, go on and on. He seems to have produced one every two or three days, year round, year after year; and each one, or a copy thereof, would be put in the mail and sent to those on his mailing list. Even though I was never one of the people to whom he sent his work—not ranking with Göttingen's Akademie der Wissenschaften and Paris's Académie des Sciences (both of which acknowledged receipt of L.'s material)—I have a stack almost four inches high of his output, dated from 1968 to 1987. The sheets are all very similar. Never is there an introduction, never an explanation: the pages all start abruptly and all end the same way. Sometimes their meaning is impenetrable and sometimes it is transparent, though their purpose has always been a mystery to me.

L. has irked me. One of his sheets ended with:

Gentlemen if your computers can give the sum of $1/N$ to 2000, I would appreciate a note telling me the total.

L. had not sent me copy of the sheet, but when one fell into my hands I wrote to L. to tell him that

$$\sum_{n=1}^{2000} \frac{1}{n} = 8.17836810361028,$$

approximately. He never acknowledged my mighty effort of computation. But another of his sheets, devoted to the equation

$$\left(\frac{1}{N} \cdot \frac{1}{(N+1)} \cdot \frac{1}{N+2} \right) + \cdots + \frac{1}{N(N+1)(N+2)} = \frac{N(N+3)}{4(N+1)(N+2)},$$

(I have copied the equation accurately) ended with:

> Professor of mathematics Underwood Dudley of DePauw, informed me

$$1 + 1/2 + 1/3 + \cdots + 1/2000 = 8.178368$$

Further, I had written to him another time asking, in what I thought was a nice way, just what it was that he thought he was up to. But there was no reply to that either. Here was a real mathematician (for all he knew) willing to correspond with him, but he declined the opportunity. He also did not put me on his mailing list. That is unusual behavior for a crank, but then L. was a very unusual crank—unique, in fact.

Second only to L. in productivity was E. O., who issued one-page *Bulletins* almost as frequently as L. produced pages. O., however, typed his work and duplicated it by mimeograph. That took time, so my stack of his material is only two inches thick. Moreover, it is less comprehensible than L.'s. Here is a sample, the *Bulletin* for August 23, 1973:

THE UNITY OF 1

It is proposed that 1 is a unity of 1 in itself but that 1 evolves as a unity of 1 in sums of 1 by ratios of sums of 1. Thus, a simple unity of 1 evolves in ratios of sums of 1 as 1 111 111 111 and a complex unity of 1 evolves in ratios of sums of 1 as $1\ 111\ 111\ 111^1$, $1\ 111\ 111\ 111^2$, and $1\ 111\ 111\ 111^3$.

It is proposed that the sum of $1\ 111\ 111\ 111^2$ evolves further as a principle of unity in a simple compact ratio defining a simple mean value of 1 in 123 in the simple ratio $(1 + 3)/2 = 2$. A constant mean value of 1 in 123 456 789 0 is proposed in which the unity of 1 is conserved by a more complex ratio:

$$\left(1^3 \times \sqrt{2} \times \frac{1^3 \times \sqrt{2}}{3^2}\right)^2 = 222\ 222\ 222\ 222^2 = 49\ 38\ 27\ 16\ 05.$$

The above ratio is proposed as the fundamental form of the complexity of nature in its utmost simplicity. As such, it is proposed as the prototype of all of nature in its numerical composition as a ratio of sums of 1 in mean constancy and in dynamic equilibrium as a system of three-body systems. The sum of 49 38 27 16 05 as a formulation of the primary integers in a reciprocal spiral of 43210 and 98765 is proposed as fundamental in nature, whether in the structure of a spiral galaxy or a sunflower.

The sum of 49 38 27 16 05 as the primary structure of nature is proposed particularly as inherent in the structural fundamental constants and in the structural wavelengths of light by the very simple ratios of 49 38 27 16 05 in which they evolve.

Another *Bulletin*:

THE MUTUALITY OF NUMBER AND NATURE

A complete mutuality of number and nature is proposed in the following formula, where number as a flow of prime numbers defines nature as a flow of particles:

$$\left(12531^3 \times \sqrt{2718} \times \frac{12531^3 \times \sqrt{2718}}{3141^2}\right)^2 = 1\ 137\ 773\ 773$$

The mutual flow of number and nature in the above formula implies a constant variability of the formula in its sums in harmony with the constant variability of nature.

The formula derives from the fundamental constants.

The left-hand side of the last equation is a recurring form in O.'s work, though why he did not simplify it to $a^{12}b^2/c^4$ is not clear.

One more example, of many possible:

THE SYMPHONY OF NUMBER

A theme with variations

It is proposed that number flows from the integers in the form of music. Thus, number flows in major and minor triads by multiple inversion, as in the following variable triad of integers:

$$468 \qquad 684 \qquad 846$$
$$486 \qquad 648 \qquad 864$$

The form of music in the above triad is proposed to be manifest in nature in the following numerical derivation and composition of the charge to mass ratio for the electron:

468	684	846
/ 266 090 090 0	/ 388 900 900 9	/ 481 009 009 0
= 175 880 2817	= 175 880 2817	= 175 880 2817

486	648	864
/ 276 324 324 3	/ 368 432 432 4	/ 491 243 243 2
= 175 880 2817	= 175 880 2817	= 175 880 2817

It is hard to know what to make of that, and of sheet after sheet of similar material. Mysticism is what I made of it, with its connections of number to music and its echoes of the Pythagorean "All is number." That is what I wrote to O., along with comments about the non-applicability of mysticism to physical reality, and its lack of association with rationality. Here is part of his reply:

> Your problem with my work is understandable but it may already be in the process of being solved with my latest paper where I "propose" that every standard wavelength of light is, or approaches, a simple harmonic ratio of 19 1326 5300. On this premise I stake the rationality of my mind. I am prepared to meet any reasonable criterion you may propose for the validity of this concept. But your best response would be to explore a few wavelengths on your own and to see what I see for yourself.
>
> I would be very happy to have you explain the sheer fact of the approach of 7771.928 as the standard value of a particular wavelength of oxygen to the following ratio:
>
> $$777\ 192\ 8086\ /\ 19\ 1326\ 53004 = 5800\ 000\ 000$$
>
> Be assured that such ratios can be adduced endlessly.

Dated just two days after that letter to me was another *Bulletin* whose first paragraph was:

> What has mathematics done to number and to the human mind? The extraordinary difficulty of seeing number in the simple harmony of its own nature makes one tremble and shudder for the soul-destroying effects of

that attitude toward and use of number that we call mathematics. The utter delight in and dependence upon artificiality and sophistication that attends mathematics and its prophets has made us purblind to the wonders and glories of natural number. It is as though we had enslaved number to do out bidding and could never dare to look number in the eye and see and respect and appreciate it in its own merit. Not that mathematics has not used number with great effect for pragmatic purposes. We can even plot our way through space! But do we know where we are going in the whole of space?

I think that I had knocked O. slightly off balance, and his *Bulletin* was part of the process he was going through to regain his equilibrium.

O. thought that the value of π could be improved on:

A THEORY OF PI

The ultimate term of relation in number is pi, with an assumed value of 3.14159..., defining the relation of the diameter and the circumference of a circle. The *primary* term of relation in number is *1*, defining a unitary relation of difference from 0. A *primary extension of 1*, in distinct multiples of 1, may be seen in the number *157* as delineated in the following number form:

$$
\begin{array}{ccccccc}
0 & & & & 6 & & \\
& 1 & & & 5 & 7 & \\
& & 2 & & 4 & & 8 \\
& & & 3 & & & 9
\end{array}
$$

The internal relatedness of a circle and pi finds a precise counterpart in the internal relatedness of the above number form:

$$
\frac{248}{248 - 157} + \frac{157}{248 + 248} + \left(\frac{157}{248 + 248}\right)^2 = 3.1420
$$

In either case the numerical value is remarkably the same, far beyond all possibility of mere coincidence.

On the basis of this identity of function it is proposed that a direct and simple relation in whole numbers must exist between 157 and pi and that such relation actually appears in pi if the value of pi may be slightly adjusted to 3.14157. With this value 157 may be seen in 314, disregarding the decimal point, as multiplied by 2 and in 157 as multiplied by 1. One may then project pi as a principle of such alteration to infinity, defining the constancy of nature:

$$3.14157314157314157\ldots$$

That is, π is rational, $3\dfrac{47191}{333333}$.

O. also produced longer papers that he submitted to mathematics journals for publication, without success. I have seen no new *Bulletins* for a long time, so O. may have given up.

PUZZLE, A

S. T.'s *Number Reduction and its Application* is mostly number mysticism, eclectically combining the *I Ching* with "Om mani padme hum," and number mysticism is not mathematical crankery. However, he started his work with:

We Breathe in a Breathing Universe

$X = (XX^2)$ is the first step in reducing decimal system numbers into "prime materia". By squaring each number, one comes always to four basic numbers: $1 - 4 - 7 - 9$.

$$X : \quad 1\ 2\ 3\ 4\ 5\ 6\ 7\ 8\ 9$$

$$X^2 : \quad 1\ 4\ 9\ 7\ 7\ 9\ 4\ 1\ 9$$

$$X^3 : \quad 1\ 7\ 9\ 4\ 4\ 9\ 7\ 1\ 9$$

$$X^4 : \quad 1\ 4\ 9\ 7\ 7\ 9\ 4\ 1\ 9$$

Analogy is made to these four numbers with the four elements and their corresponding colors:

1—fire—red   9—earth/sun—yellow
4—air —blue   7—water   —green

This analogy is applied in the two colored 3×3 squares:

X^2—earth breathes out water flanked by fire and air.

X^4—earth breathes in air flanked by fire and water.

He continued, at length, in that typical mystical style.

The puzzle is, how did T. calculate X^2? If you or I were to calculate the square of 123,456,789, we would get 15,241,578,750,190,521, or something close to that, rather than 149,779,419. We would not get his numbers for X^3 or X^4 either. How did he get them?

The answer will appear later on.

PYTHAGOREANS, NEO-

In 1968, the *Mathematics Teacher* published Figure 35 under the title "Pythagorean Brotherhood Lemma," with the caption

$$A + B < = > C \text{ according as } \theta < = > 90°.$$

You can see why it was published: it is a pretty picture, and it is a good exercise for teachers and students to verify that the statement in the caption is true. However, the title should have been the tipoff that a crank was at work, and the editor of the *Mathematics Teacher* might have guessed that its author, R. S., was going to refer to his publication in that journal ever after as a seal of approval on his work. S. indeed kept referring to it. In 1987 he was still promoting his Brotherhood, and he may be at it yet.

He went as far as to insert an advertisement in the *Mathematics Teacher*, and perhaps elsewhere:

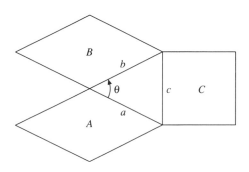

FIGURE 35

271

PYTHAGOREAN SOCIETY

$$\frac{x^2}{4^2} + \frac{(y \pm 3)^2}{3^2} = 1$$

$$x = 0 \qquad |y| < 8$$

HELP RESURRECT PYTHAGOREANS! Send return stamped envelope to [S.], [at his address], creator of the geometric figure as published in the February 1968 issue of this journal under the title "PYTHAGOREAN BROTHERHOOD LEMMA."

The graphs of the two ellipses whose equations are in the advertisement make a squashed figure eight, but what the extra conditions on x and y mean is a Pythagorean mystery.

Sending the stamped envelope to S. would probably have gotten you a copy of the *Pythagorean Society Charter*, headed with the same unexplained equations. The Charter stated:

> Axiom: The spirit of Pythagoras by the power of the seven spirits of God led me to find the geometric figure as published in Feb 68 Mathematics Teacher journal under the title Pythagorean Brotherhood Lemma. Picture a right triangle with squares on its sides. Keeping the lines forming the right angle of said right triangle straight, modify such with help of conjecture by having said right angle become acute (obtuse) enough so that one can see the living truth of mathematics. Color figure black on a red background all within a gold ring. Said figure is on one side of the great seal of the Pythagoreans.

The "conjecture" mentioned there is explained later:

> Conjecture found by myself using the Pythagorean method of correspondences led by the seven spirits of God. Revelation 11:1 .. Measure calls one's eyes to a picture of a right triangle with squares on its sides. See the other big (equal) square and the rod (straight line) through the 3 points of the 2 little squares (or parallelograms with equal sides for each in axiom). Omit the 2 big (maybe unequal) squares and see a rod like to a reed. Let 2 little squares (maybe parallelograms) become like to one's eyes and see a reed like to a rod. Symbol thus found defines the parameters for its equation (top of Charter). Colors connected with God's throne show color of symbol to be green (bottom eye), red (top eye), & blue (rod) on a white background all within a gold ring. Said symbol slightly modified with eyes slightly closed using rectangles is on one side of the great seal

of the Pythagoreans. Picture (reed = throne = lamb = name / 11:1; 4:6; 5:6; 3:12; 14:1)! See the same picture in the 2 seas and river of Israel. See how the moon and sun make an exact fit during a total black-out of the sun as seen by man (woman) on earth. See how both the moon and sun follow the path of the milky way across the heavens. One may come to see that God put said signature in our sky for all to see. Acts 2:21 shows importance of God's name.

I have quoted at length because S. is an example of the mathematical-religious mystic, and his writing is typical. Similar examples appear in De Morgan's *Budget of Paradoxes*, published in 1872, showing that the type of mind that inclines that way continues to reappear, generation after generation. It may seem strange that Pythagoreanism should get mixed with Christianity, but the gaudy imagery of Revelation has an irresistible appeal to some mystics.

The Charter concludes with:

Purpose: Pursuit of the mathematical way of life endowed with religious significance by the seven spirits of God.

Organization: Secret and known only to the seven spirits of God.

Membership: Confirm such by using finger to sketch symbol in order of bottom eye, top eye, and rod on the part of the forehead (space above the eyes) located between the eyes. As such is being done, think on the rest of the rod (the seven spirits of God) being within us Pythagoreans.

I doubt if there were many of "us Pythagoreans."

The 1987 letter by S. that I saw was wild and incoherent, and had much in common with communications whose return addresses are hospitals.

PYTHAGOREANS, THE MYSTERY OF THE

If you look in any history of mathematics you will learn that the Pythagoreans were a religious brotherhood that flourished in ancient Greece in the sixth and fifth centuries B.C., founded by Pythagoras of Samos after whom the Pythagorean theorem is named; that they taught that all is number; that 10 was holy; that they discovered that $\sqrt{2}$ was irrational; that they noticed that musical tones sound well together if their frequencies have ratios consisting of small whole numbers; that they believed in the transmigration of souls; that they classified numbers as square, triangular, perfect, and so on; that they ate no beans; and that they withered away.

That is all very well, but it leaves several questions unanswered. The Greek mathematical tradition was geometric, not arithmetic as was the Pythagoreans'. Greek philosophy was rational and had no nonsense about beans in it. The scholarly works of the Greeks did not wither away and are still read today. The Pythagoreans were *strange*. What about them?

J. B. provided an answer in his 28-page pamphlet *Babylonian Arithmetic*. As the title shows, his purpose was not to explain the Pythagoreans, but to promote the system of arithmetic used by the ancient Babylonians in the land between the Tigris and Euphrates rivers, millennia ago.

The Babylonians counted by 60s, and parts of the Babylonian sexagesimal system survive in how we divide hours and degrees into minutes and seconds. The purpose of B.'s pamphlet is to revive the Babylonian system:

> The Romans had a motto—to the stars through difficulties. We use special-type, sexagesimal numbers to measure time and angles. The Babylonians used not-quite-the-same six-by-ten numbers for all of their counting and measuring. They had to use a six-by-ten arithmetic to fit their

six-by-ten numbers. It is nearly as easy to use as decimal arithmetic; it is more difficult to learn to use. It was the arithmetic of our past; it will be the arithmetic of our future. You can learn to use that arithmetic. When you move through that little difficulty, you will not move yourself all of the way to the stars. You will move our whole human race one small step closer to those distant stars.

The Babylonian system has superiorities, B. said:

Americans learn in grade school that hundred-digit numbers and thousand-digit numbers are logical possibilities. Most American children do not produce twelve ten-digit numbers during twelve years in grade school and high school. There are nineteen numbers [on a Babylonian clay tablet] that contain ten or more digits. Neugebauer reported a Babylonian number on another inscription that contained more than thirty digits. He reported dozens of numbers that contained more than twenty digits. Babylonian children learned to produce correct values of twenty-digit numbers and even thirty-digit numbers as a matter of grade-school routine. At least in that respect, Babylonian arithmetic four thousand years ago was substantially superior to American arithmetic today.

The ancient Babylonians wrote on wet clay; paper, an abundance of which we take for granted, lay thousands of years in their future. Wet clay is not the handiest medium to use for doing calculations, so it is very probable that an ancient Babylonian calculator used something like an abacus to carry out arithmetical operations. The mechanical aid did not have to be like a modern abacus, with beads sliding on wires: pebbles in grooves or discs on lines would serve as well. B. was convinced that there were ancient Babylonians who made their living as abacus-operators, and also that their use of the abacus explained their lack of a symbol for zero:

Babylonians did not write zero digits and decimal points in their inscriptions. That omission often is mentioned as the "proof" that Babylonians did not understand the mental concepts that we associate with visible zero digits and visible decimal places. The term "place value" has more "place" in its meaning when a number appears on an abacus than when it appears as digit symbols written on paper. An abacus automatically presents a zero digit at every place where any value might be indicated unless an alternative digit symbol is put at that place. When the computing is done on an abacus, there is much less need to put explicit zero symbols into written records. The omission of such symbols from such records helps keep the professional skill of the professional abacus operator concealed from the public. Even though Babylonians did not write explicit zero-symbols in their inscriptions, they had to develop clear mental conceptions in order to put their symbols for not-zero digit values at the right places on abacuses. The real issue here is not the amount or the quality of the evidence that Babylonian inscriptions supply; it is the earnestness of the European determination to refuse to concede Asiatic Babylonian intellectual competence.

B. had taught himself how to use a soroban, the Japanese version of the abacus, and was a firm advocate of its use:

> This author assumes that Babylonian parents lacked the technology to build sorobans four thousand years ago; hence, Babylonian children did not learn to operate sorobans then. They did learn to operate abacuses of some not-identified type. Now that this author has learned soroban operation, he can perceive positively that Babylonian inscriptions define drills in abacus operation. . . . Babylonian children four thousand years ago developed a competence in arithmetic that American children do not develop now. We assert that our American schools should teach abacus operation and six-by-ten arithmetic today so that American children today can develop that long-ago competence in arithmetic.

When B. said that the Babylonians used something analogous to the abacus for calculations, he was not departing very far from the general opinion of those people who think about such things. When he said that schools today should invest in sorobans for every pupil he crossed the line into crankery.

His explanation of why the Pythagoreans were the way they were crossed the line back into historical speculation:

> If any American today tries to learn what made Greek Pythagoreans Pythagorean, he finds a baffling lack of agreement in different encyclopaedia articles. The Pythagoreans were a Greek cult with a not-Greek cult policy—they refused to enter into public discussions of cult doctrine. They built no temples and supported no priesthood. They had a special interest in numbers and perpetuated an extensive amount of information about numbers. They maintained a quasi-scientific conviction that all things in nature might be explained by use of numbers. They never explained their doctrines themselves; all of the comments about Pythagoreans and their doctrines were written by not-Pythagoreans, some of whom claimed to know Pythagoreans.

> We deny that the discussions of Pythagoreans and their doctrine in present-day encyclopaedias really add up to coherent good sense. Neugebauer has provided a convincing demonstration that Pythagorean number

theory was known in Babylonian territory long before the Greeks learned to use the alphabet. Surely, Greek Pythagoreans did not invent that theory. There are humans alive today who are named "Aristotle". We do not expect them to be sources of Aristotelian philosophy. We guess that the Greek Pythagoras was named even more years after all of the Pythagorean doctrine was first taught in Babylonian territory. That gives us room to put some coherence into Pythagorean doctrine and character.

Greek Pythagoreans were, so we say, immigrant abacus operators. They had gained the Babylonian equivalents of American college educations in engineering, accounting, and other subjects relevant to number-processing. They identified themselves as preservers and expanders of packages of knowledge and skill that a Master Teacher had imparted to the human race. Babylonian culture had provided traditional government subsidies that had sustained those packages of knowledge and skill as family-heirloom occupations. Greek culture was thoroughly committed to the use of decimal arithmetic. It did not provide subsidies to sustain six-by-ten arithmetic and other not-immediately-useful components of Pythagorean knowledge. At least some of the immigrant Pythagoreans found employment as abacus operators. Pythagorean fathers could command their sons in Greece to learn six-by-ten arithmetic and other component parts of Pythagorean doctrine as children: Pythagorean grandfathers could not persuade adult sons to command child grandsons to learn Babylonian mumbo-jumbo that would produce neither respect nor income within Greek culture. Pythagoreans in Greece could form associations like medical doctors and plumbers in America now form associations; those long-ago associations had no more desire to build temples and support priesthoods than present-day associations of professionals do. Grandsons willingly would learn the income-producing skill of abacus operation that grandfathers and fathers practiced; otherwise, they wanted to be as Greek as the Greeks around them. After grandfathers died, Greek grandsons felt a Greek freedom to put scraps of the teachings of their grandfathers into Greek writing—at least as long as their discussions of those teaching would not be technical explanations of abacus-operation techniques. Our explanation accounts for the known Pythagorean interest in numbers.

It does indeed. B. went on to explain that the Pythagoreans left Babylon because Nebuchadnezzar cut off their subsidy and began to persecute them. B.'s reading of history also explains why the Pythagoreans died out: succeeding generations of hereditary abacus-operators became assimilated into Greek culture and put aside the alien superstitions of their ancestors. He did not speculate that the immigration of the Pythagoreans to Greece was like the immigration of German scientists to the United States in the 1930: that immigration stimulated American science and mathematics so tremendously that its effects are still being felt, and that it is not farfetched to think that the Pythagorean immigration had similar effects, if not greater ones. Perhaps the glory that was Greece had its origins in ships sailing west from what is now the coast of Syria, which had as passengers Babylonian abacus-operators and their families fleeing their native land, but taking their abacuses with them.

Well-written speculation such as B.'s is delightful. However, crankery crept back in when he went on to explain other things:

> In the opinion of this author, the Bible story of Adam and Eve is a report about a two-person colony from another solar system that was installed on this planet. The Garden of Eden was the school for colonists that they attended. They entered that school, Adam first, as bawling babies perhaps five years apart, naked and not ashamed. They left as college-educated newlyweds, dressed in new wedding garments; they had finished the layout (= plan, curriculum) of the knowledge of good and evil. The Hebrew Bible preserves an important share of their training in theology, ethics, government, and religious ritual (subjects of interest to Moses). It omits virtually all of their training in mathematics, science, and improvements of technology (subjects of special interest to this author). It is remarkably free from anti-science. Its discussion of the days of Creation provides some helpful hints about the technology of installation of a colony of humans in another solar system.

B. had a purpose for beginning his pamphlet with a reference to *ad astra per aspera*. The stars were where we came from, and to them we should return. Learning to use the abacus is a first step.

QUADRATURE OF THE CIRCLE

There have always been circles. The circular disk of the sun, the circle of the full moon, the circle of the horizon: the human race has always been aware of circles.

The race has an insatiable curiosity about almost anything, and circles are no exception. Sometime, long ago, someone realized that if one circle is three times as wide as another, it is also three times as far around. Of course, we think, that is obvious and not a large discovery. But it *was* a large discovery. Most of us think it is obvious because we have been *told* that it is obvious, told enough times that it has been filed in our minds with all of the other unquestioned rules to which we accede without thinking about them: the earth goes around the sun, hot air rises, electricity makes magnetism. The discovery of proportion takes some mental maturity. Give a young child this problem:

> Robbie can write 5 words in the same time that Annie can write 3.
> How many words can Robbie write in the time that Annie can write 15?

You may get the right answer, but there is a good chance that you will get the firm reply, "17." Like the invention of the wheel (another circle), the discovery of proportion is taken too much for granted.

Once the idea of proportion was part of the intellectual heritage of the race, the application of it to circles was natural. The ratio

$$\frac{\text{circumference}}{\text{diameter}}$$

is the same, no matter what size the circles are; so the question arises, what is its value? The question was asked by all of the ancient civilizations, and answered by

279

them in one way or another. Even though most ancient civilizations were very practical, it could be argued that determining the ratio is a question of pure mathematics,
asked for its intellectual interest, since circles do not often present themselves in
nature as things to be measured. Of course, it is possible that the question arose
when one of our distant ancestors dug a circular hole, for some reason wanted to
know its circumference, and inquired of the local person known to be learned in the
lore of quantity. "Ah," that person might have said, "that is a new idea. Let me think
about it." However, I like to think that the nature of the circle was first investigated
for its own sake, and only later were the results of the investigation applied to holes.

The ancient civilizations in the Middle East and China viewed mathematics as
an experimental science, so experiment was the way to find the value of the ratio.
We do not know the nature of the experiments—perhaps ropes made to follow the
outline of a circle, maybe wheels rolled on the ground—but we have the values
that were found. The ancient Egyptians, whose mathematics was fairly crude, hit
on $(16/9)^2$ as the value. That is $3.16049\ldots$ and is not crude, being in error by
only six-tenths of one percent. It would be perfectly acceptable for any practical
use to which it was put. Later we will see that a nineteenth-century artisan used
$3\frac{1}{6}$, not as accurate as $(16/9)^2$, and found that it admirably fit his work. The ancient
Babylonians seemed to have used $3\frac{1}{8}$, also a close approximation. The Babylonian
value was too small and the Egyptian one was too large; it is too bad that no one
thought to synthesize and take the average:

$$\frac{(25/8) + (16/9)^2}{2} = 3.142747\ldots$$

would be a value even better than $22/7$. (The value $22/7$ did not occur to the ancients,
probably because their systems of calculation made sevenths difficult to deal with.
Calculating with the Babylonian or the Egyptian value involves only divisions by 2s
and 3s.) A variety of values for the ratio are found in old Chinese works, but what
value was used in the dawn of Chinese civilization is unknown because no written
records survive from that time.

It is probably time to stop being coy and call the ratio by its proper name, π.
The reason for avoiding its use is that the ancients did not use a symbol for the
ratio. They used words instead, and the use of π did not become standard until Euler
started calling the ratio by that name in the eighteenth century. It is a good notation:
π is easily recognizable; is not used for anything else in elementary mathematics;
and it has meaning, π standing for the first letter of "perimeter" no matter whether
the word is written in Greek or English.

The ancient Greeks changed geometry from an experimental science to a
deductive one, thus changing the course of mathematical history. They were not

particularly concerned with the numerical value of π or with the numerical values of anything else—those could be left to people engaged in business and other vulgar trades—but they did try to find a square with the same area as that of a circle using only the tools of Euclidean geometry, namely straightedge and compass. They did not succeed, and they probably concluded that any such construction was impossible. They did succeed at squaring the circle using more than straightedge and compass; but these achievements were geometrical and not numerical, so did nothing toward determining the value of π.

What the geometrical constructions do illustrate, besides squares with the same areas as circles, are the amazing intellectual accomplishments of the ancient Greeks. Before 400 B.C., Hippias of Elis had constructed his quadratrix, a curve that can be used to square the circle. He did this hardly more than 400 years after the human race had awoken from its millions of years of sleep to the dawn of reason. Who Hippias was and how he came to think of his quadratrix we will never know, but we do know how to make the curve.

Take a quarter-circle (Figure 36). Start two points moving down from A at time $t = 0$ at constant speeds, one along AO and the other along the arc of the circle, so that both will reach the x axis at the same time, say when $t = 1$. At time t, the point moving down AO has reached some point, call it C, and the point moving along the circular arc has reached another point, D. Draw a horizontal line through C and draw a line from O to D. Where they intersect, at Q, is a point on the quadratrix. The collection of all such points for all values of t, $0 \leq t \leq 1$, is the quadratrix.

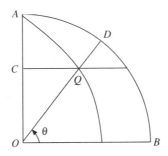

To see how it squares the circle, let us derive its parametric equations. Hippias of course did no such thing since the discovery of analytic geometry lay more than two thousand years in his future, but because Hippias did not have a tool is no reason for us not to use it. As C slides down the axis from $t = 0$ to $t = 1$, it will be at $(0, 1 - t)$ at time t. As D slides down the arc, the angle θ will be $\frac{\pi}{2}(1 - t)$ at time t. So, if the coordinates of Q are (x, y), then

$$\frac{y}{x} = \tan\theta,$$

or, since $y = 1 - t$,

$$x = \frac{y}{\tan\theta} = \frac{1-t}{\tan\frac{\pi}{2}(1-t)}.$$

To see where the quadratrix meets the x axis, take the limit as t goes to 1:

$$\lim_{t\to 1}\frac{1-t}{\tan\frac{\pi}{2}(1-t)} = \frac{2}{\pi}\lim_{t\to 1}\frac{\frac{\pi}{2}(1-t)}{\sin\frac{\pi}{2}(1-t)}\cdot\cos\frac{\pi}{2}(1-t) = \frac{2}{\pi}.$$

Thus, the distance from O to B is $2/\pi$, and it is not difficult to get from that a line segment with length π, from that a line segment with length $\sqrt{\pi}$, and from that a square with the same area as a circle with radius 1.

Another curve with which the circle can be squared is the Spiral of Archimedes (Figure 37), whose equation in polar coordinates is $r = \theta$. The spiral of Archimedes is like the quadratrix of Hippias in being generated by two moving things: the radius OA, rotating at a constant rate; and the point R, moving out along the radius at a constant rate and tracing the spiral. Greek geometry is often thought of as static, but inside the minds of the Greek geometers points and lines were moving. Given the spiral, if θ and R are moving so that $\theta = kt$ and $|OR| = kt$ for the same constant k, then the length OB is π and the circle can be squared.

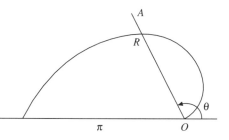

FIGURE 37

Archimedes was much less pure than the other Greek geometers of the golden age. That does not mean that he was corrupt, but that he did not mind turning his hand to practical problems. He found a way to tell if the king's crown was pure gold without snipping off a piece, he (allegedly) burned Roman ships using mirrors and the sun, he estimated how many grains of sand it would take to fill the universe, and his pulleys helped move things that were otherwise immovable. He not only showed how to find a line segment of length π with his spiral, he found numerical limits between which the value of π had to lie. His idea was the natural one (easily seen to be natural after it has been thought of; getting a "natural" idea in the first place is another matter entirely) of inscribing and circumscribing polygons in and around a circle, so that the circumference of the circle is bounded above and below by the perimeters of the polygons. Any student of mathematics ought to be able to apply the idea with squares (Figure 38) and find that, if the radius of the circle is r, then the side of the circumscribed square is $2r$ and the side of the inscribed square is $r\sqrt{2}$. Thus the circumference of the circle is larger than $4\sqrt{2}r$ and less than $8r$, so

$$2\sqrt{2} < \pi < 4.$$

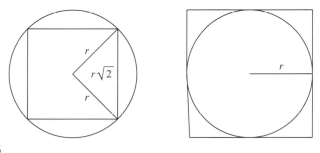

FIGURE 38

Those bounds are not very close, but then the squares are not very close to the circle. Hexagons will give closer bounds, and using regular polygons with ninety-six sides, as Archimedes did, will give bounds that are close indeed:

$$3\frac{10}{71} < \pi < 3\frac{1}{7}$$

or $3.140845\ldots < \pi < 3.142857\ldots$. The average of those two values would be a super approximation to π,

$$3\frac{141}{994} = 3.141851\ldots,$$

but Archimedes was not concerned with accuracy of approximation. The idea was what was important. Archimedes's idea was still being used a thousand years later to find approximations for π.

Of course, finding approximations to π is not the same as squaring the circle. Even though the ancient Greeks found many ways to square circles (using more than straightedge and compass, of course), people will not stop trying to square circles. Their efforts fall into two classes, geometrical and numerical. The geometrical circle-squarers produce straightedge and compass constructions and claim that they square the circle perfectly. The constructions of necessity produce values of π different from $3.1415926535\ldots$, but pointing this out usually has no effect on the circle-squarers, who usually respond by saying that the construction is geometric, made entirely of lines and circles, and numbers have nothing to do with it. The numerical circle-squarers do not bother with a geometric construction, but merely produce a value of π from somewhere or other, with more or less reasoning (sometimes none) to justify it.

The numerical circle-squarers who maintain that $\pi = 3\frac{1}{8}$ (a popular value) can be told that Lambert proved in 1761 that π is irrational and hence cannot be $3\frac{1}{8}$. And those who assert that $\pi = \sqrt{10}$ (another popular value) can be told that Lindemann proved in 1882 that π is transcendental and hence cannot be $\sqrt{10}$. They can be told this, but it probably will not do any good.

Even after Lambert showed that π is irrational, there remained the possibility that a geometric construction might square the circle, producing $\sqrt{\pi}$ exactly with straightedge and compass. After Lindemann's proof, that possibility disappeared since transcendental numbers cannot be constructed with straightedge and compass alone. Circle-squarers can be told this as well, but since most do not know about transcendental numbers, it will bounce off.

Even after hopes for squaring the circle died among the non-crank population, interest in π did not disappear. With the revival of learning in Europe came the revival of interest in π. Viète in 1593 gave the formula

$$\frac{2}{\pi} = \sqrt{\frac{1}{2}} \cdot \sqrt{\frac{1}{2} + \frac{1}{2}\sqrt{\frac{1}{2}}} \cdot \sqrt{\frac{1}{2} + \frac{1}{2}\sqrt{\frac{1}{2} + \frac{1}{2}\sqrt{\frac{1}{2}}}} \cdots,$$

exact, but not suited for computation. Van Cuelen calculated π to more than 30 places in 1596, and progress has been made ever since. Currently, π is known to more digits than anyone would ever need. I would quote an exact figure, but the mathematical subfield of π-calculation has been making extremely rapid strides recently, old records are frequently being broken, and a new one may be set any day.

So, as far as mathematics is concerned, the question of squaring the circle is settled. It is possible to square the circle geometrically, using more than Euclidean tools, and it is possible to square the circle practically, calculating its area by multiplying π by the square of its radius. The more accurately the value of π is known, the more accurately the circle can be squared. Since π is now known to more than a billion digits, we are able to square the circle very nicely, thank you. (A billion digits are not really necessary. Since the constants of nature—the speed of light, the acceleration of gravity, and so on—are known to only eight or nine significant figures, eight or nine digits of π are enough for any practical purpose, and that many have been known for a long time.)

The first 314 digits of π, in case you need them, are

3.14159 26535 89793 23846 26433 83279 50288 41971 69399 37510
58209 74944 59230 78164 06286 20899 86280 34825 34211 70679
82148 08651 32823 06647 09384 46095 50582 23172 53594 08128
48111 74502 84102 70193 85211 05559 64462 29489 54930 38196
44288 10975 66593 34461 28475 64823 37867 83165 27120 19091
45648 56692 34603 48610 45432 66482 13393 60726 02491 41273
72458 70066 063

You may think that there is nothing to be gotten from a listing of digits, but it contains surprising things. For one, the best approximations to π with fractions of two and three digits are $22/7$ and $355/113$. The 7th, 22nd, 113th, and 355th digits of π are all 2. There is symmetry around the 22nd digit of π: 46 2 64. in addition, 462 and 264 are both multiples of 22. Looking at the digits around the 22nd in a different way,

79 32 38 46 26 43 38 32 79,

there is symmetry again. 79 is the 22nd prime. 79 is also the sum of the divisors of 365, the number in the year's circle of days. Around the 79th digit of π is another example of symmetry:

628 620 8 998 628.

628 are the first three digits of 2π. In addition, $620+998 = 1618$, the first four digits in ϕ, the golden mean. π has depths.

What follows is a sampling of the work of circle-squarers, both geometrical and numerical, most recent but some old, some bad and some good (as circle-squarers' work goes). It is meant to be representative of the field.

An example of the practical circle-squarer, the person who thinks that π is not known well enough for everyday use, is found in *The Measure of the Circle*, a 155-page book published in Rhode Island in 1845. Its author, J. D., "discovered"—how he does not say—that π is $3\frac{1}{6}$ exactly. D. wrote:

My experience as a mechanic has taught me the use and importance
of this measure. I have been engaged and employed as a superintendent,

having the care and sole management of cotton, woolen, and silk machinery, and have gained the approbation of my employers, my management
being such as to save them, yearly, a large amount. They said they could
not comprehend how I could manage to such perfection; for when I had
perfected a thing, it was sure to answer the purpose. This perfection I
arrived at by the measure of the circle, unknown to them.

The expression was made use of one of my employers, in a cotton
manufacturing village, "How is it that it does not cost you one half to
keep your machinery in order that it does the rest of us?" The answer was,
"When [D.] calculates, it is sure to come right; while others have to do
their work five or six times over."

That tells us something about the level of arithmetical ability in New England one
hundred and fifty years ago.

The book is full of information, including:

A gentleman by the name of North, a descendant from Lord North,
of England, contemplates building a glass globe, one mile in diameter, in
the United States, with the principal places in the world painted within
the globe, so that, by the construction of seats within, the people can sit
with ease, and see the operation of the earth, and all its principal places.
How many square feet of glass would it require?

It ends with some testimonials. One is:

I have travelled with Mr. [D.], in England and the United States, for
six years; we have visited all the most learned mathematicians that we
could hear of, but have never found one that attempted a disapproval of
his work. I have examined, to the best of my ability, the works of writers
on mathematics since the time of Euclid; and without a doubt, according
to mathematics, he has solved the remarkable problem, surprising as it
may appear to many.

That is remarkable. Who were the learned mathematicians? What did they say?
Were D. and his companion listening when they were told that $\pi = 3.14159\ldots$?
What works did the writer of the testimonial examine? Did he mean everything he
wrote? There are mysteries there.

Another testimonial is less mysterious, and shows why the ancients had no
need for approximations of π better than $(16/9)^2$ or $3\frac{1}{8}$:

I have wrought as a mechanic for twenty years, and in some of my
mechanical operations I have found it very difficult to match my work
from the proportion of as 7 to 22, and by experimental operations I came

to the measure of three times the diameter, and one sixth, and from this
I have found no difficulty in matching my work; and when Mr. [D.] told
me that three and one-sixth times the diameter was his proportion, I was
satisfied that his measure was correct.

In addition, a "Civil and Military Engineer" said:

I have examined the measure of the circle by [D.], and beyond all
doubt it is perfect measure.

And, worse yet, a "Professor of Mathematics" was quoted as saying:

I have examined the measure of the circle by Mr. [D.], and find it,
in my opinion, a most complete, scientific, mathematical measure of the
circle.

It is very curious. Surely in 1845 it was generally known, especially by professors
of mathematics, that π was 3.14159... and that value must have appeared in all the
appropriate textbooks, but D. never mentioned it. As far as a reader of his book can
tell, the value of π was, before D. found it, completely unknown.

Since J. D. was an engineer, he probably found his value of π by mechanical
means. Teachers of mathematics should make it plain that mathematics is not a
branch of engineering and that mathematical truths cannot be found by manipulating
physical objects. Most probably they do try to make it plain, but the message does
not always get through.

C. D., the author of *Mathematical Commensuration*, a 31-page pamphlet published in Illinois in 1883, did not get the message. D., who put "D.Ph." after his
name, decided to apply the principles of "object-teaching" to geometry. So,

a great number of wooden cubes were provided.

D. then laid them out:

By having several hundreds of these cubes placed in succession, it
was noticed that every 17 diagonals equaled in length 24 of the sides, and
that, therefore, the ratio 24/17 suggested itself.

This of course was of little account, but it was a clue.

D. was not aware that his discovery that $\sqrt{2} = 24/17$ was in conflict with the
Pythagoreans' discovery, more than two millennia earlier, that $\sqrt{2}$ is not a rational
number. (Squaring both sides of his equality gives $2 = 576/289$, from which it
follows that $578 = 576$, among other things.) Encouraged, D. went on to the circle:

The next obvious experiment was to find a possible clue to the ratio
of diameter and circumference.

A number of plastic uniform cubes were formed. From some of these a number of globes were made to be sure that the same area was contained in each of the differently formed quantities. These globes and cubes were likewise placed parallel to each other,

and, by experiment, D. discovered that π was $908/289 = 3.1418685\ldots$ exactly. ("Plastic" had for D. the 1883 meaning of "deformable.") The remainder of his pamphlet consisted of arguments in favor of the new values, and it ended with:

It is not impossible that this first presentation of commensuration applied to the most profound of sciences will lead to intellectual revolution in other directions and set people thinking over a possible equitable relationship in human affairs by applying the principles of commensuration to just debate on moral, social, civil and political reform.

While it is true that making moral, social, civil, and political matters commensurable —that is, taking the irrational out of them—would be a step forward, fiddling with the value of π is not the way to do it.

The nineteenth century was, for some reason, a boom time for circle-squarers. De Morgan's *A Budget of Paradoxes* contains many quadratures, but no angle trisections and no proofs of Fermat's Last Theorem. In 1888, S. C. Gould published *A Bibliography of the Polemic Problem: What is the Value of* π, a pamphlet that contained more than 60 entries. One of them was *Quadrature of the Circle* by J. P., published in 1874 by a company that is still in business today, still publishing widely in mathematics and the sciences, and still charging prices as terrific as was the $20 they wanted for *Locomotive Engineering and the Mechanism of Railways*, by Zerah Colburn, in 1874. In the nineteenth-century style, the book's title page tells, in detail, what it contains:

Quadrature of the Circle.

Containing Demonstrations of
The Errors of Geometers in Finding
the Approximations in Use;
With an Appendix,

and

Practical Questions on the Quadrature,
Applied to the Astronomical Circles.

To Which are Added
Lectures on
Polar Magnetism,

and

Non-Existence of Projectile Forces in Nature.

The Appendix contains texts of lectures, the one on the quadrature starting:

I am here to-night for the purpose of placing before you some new views regarding the old and long-since exploded question of the quadrature of the circle. The importance to astronomy and navigation of a correct knowledge of the circle, is my reason for stepping out of my proper sphere of business, temporarily, to become a public lecturer.

Nowadays no one would turn out for a lecture on π. Times change, sometimes for the better and sometimes not.

The book was a second edition:

This work has been written several years, and was first published in 1851, and a copy sent to the principal colleges in the country. It now appears almost exactly as originally written for publication.

The author was quite aware that other writers thought that π was $3.1415926535\ldots$, but he thought that his work refuted them. It is doubtful that he convinced many other people since his book contained such theorems as:

Proposition III

The circle is the natural beginning of all area, and the square being made so in mathematical science, is artificial and arbitrary.

His value for π was $20612/6561$, or $3.1415942\ldots$; close, but not as close as the simpler $355/113 = 3.1415929\ldots$.

Another nineteenth-century circle-squarer, E. M., who wrote *Geometrical Science* in 1890, was unaware that the quadrature had been proved to be impossible:

The world has accepted the statement "That the circle cannot be squared." If some inquiring mind be tempted to ask, why? an answer is not forthcoming. The most that can be said is, that thousands have tried to solve the problem and the thousands have failed.

He also misunderstood why his value of π was not accepted:

After fourteen years hard study the author submits this first part of his work to the public. While aware that in so doing he is breasting the waves of public prejudice, and even arousing animosity, his earnest conviction that he is adding a stone to the temple whose foundation was laid by Euclid, is sufficient to animate him with the hope that time will vindicate his labors, and the domain of geometrical science attain far broader dimensions, with results, as yet, uncalculable. Of the animosity he thus invokes, the author has already received abundant proof. A carefully prepared statement of his claims, with demonstrations, has been vainly offered to the leading journals, and in all cases declined. It is humiliating to record the reasons given, which were usually as follows, viz: "general opinion is so firmly seated against any solution of the Problem, that we are unwilling to accept it, even as a communication, and fly thus directly in its face." Others, equally candid, replied: "If your figures are correct, then those of all our text-books are wrong, and we cannot afford to affront either those who publish or use them."

The idea that mathematical statements are matters of opinion or controversy is still found among cranks and hence must be alive in the population generally. How it survives is a puzzle, since teachers of mathematics do not propagate it. M.'s opinion was that $\pi = (16/9)^2$, the same value used by the ancient Egyptians.

Some modern circle-squarers, like some modern fermatists, angle-trisectors, and other varieties of cranks, try to get themselves in the newspapers. R. E. was the subject of an article in the Kansas City *Times* in 1983:

> Night has long since fallen over Kansas City, Kan., and the city is still—except maybe for the hum of trucks on the interstate, the rattle of freight cars in the rail yards and the scratch of [R. E.]'s pencil on paper.
>
> Hour by hour he labors on, alone in a house with flaking paint. Mr. [E.] stays up late looking for the nub of reality, and he is not looking in any of the usual places. Mr. [E.] believes that truth lies in the geometric concept known as *pi*.
>
> "Understand pi and you understand reality," said Mr. [E.], a 39-year-old railway clerk who points out that Albert Einstein, too, worked as a clerk before he was recognized as a genius.
>
> And Mr. [E.] is sure he has cracked it—just as sure as the experts are sure that Mr. [E.] is a serious example of a mathematical crank. Cranks are the math world's equivalent of the backyard inventor—but with a difference. Their inventions, the math establishment says, never work. . . .
>
> Mr. [E.] laughs at such criticism. "People can scoff: that's all right with me," he said, "They'll learn soon enough, I guess. I don't think anybody can deny what I've done. If they do, they're a fool. . . ."
>
> "What I've done has eluded every mathematician for 2,000 years," he said. "If I'm right it would simplify our mathematics to the point where a child would understand everything there is to understand by the time he's 10 or 12 years old—once this is standardized."
>
> That may be awhile, although Mr. [E.] has mailed copies to Harvard, Yale, Princeton, the University of California and assorted officials. His congressman, Rep. Larry Winn Jr., wrote, "Your work does seem to carry

some significance." But bastions of learning have been loath to recognize his achievement.

"It's hokum," said Paul Leibnitz, chairman of the University of Missouri-Kansas City math department. "I can't make head or tails of it."

E.'s work is all mixed up with physics and falls into the utterly incomprehensible class. One of his comprehensible accomplishments was:

> I have succeeded in the tri-section of the line segment using only compass and straightedge.

E. did not know the problem had already been solved. His value of π is variable: $\pi = 3 + 1/R$ and "varies as acceleration." It is not clear what R is. In another place he wrote that $\pi = 47/15$. Of course,

> He said he doesn't care that mathematicians call him a crank or see no problem with pi as it stands. "In my own mind I know I've got something too good to pass up," he said. But he wishes the world was more open-minded—for its own good.
>
> "I'll be done with my new diagram pretty soon, and I'll be sending it out," Mr. [E.] said, "and any mathematician who can't see what I've done is not a mathematician."

He sent it out, but not much came back. In a letter, E. wrote:

> I sincerely appreciate your reply to my last inquiry concerning pi. You are the only person as yet to even consider my work and give a direct reply.

His letter concluded:

> The danger to the life of this planet is enormous! If my discovery is correct, the President has allowed a national election to proceed without this being made public. I consider that at least dangerous, at most treasonous.

Besides trisecting the line segment, he also trisected the angle. His construction for that is at least clear (Figure 39): take the angle BOA, bisect it with OC, draw an arc with center at C and radius $|AC|$ to locate E, draw AD perpendicular to AB with $|AD| = |AB|$, and then connect D and E to locate the trisection point T. It is a good exercise in trigonometry to find how accurate the trisection is for, say, a $60°$ angle. It is a harder exercise, though still good, to find where T is for a general angle θ. Providing these exercises is the only value E.'s work will ever have.

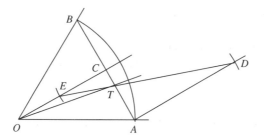

FIGURE 39

A circle-squarer both typical and colorful was L. C., who published *[C.'s]*
Unique Mathematical Geometrical Findings in 1967, with a reissue in 1973. My
copy of the first edition of this 18-page booklet has a picture stapled to its front
cover, probably of where C. lived. In the background is a grassy field, trees, and a
white fence; in the foreground is a post to which nine signs are affixed. One of the
signs reads

DENIAL
IS A
LIE!

Another,

APATHY
IS
IDIOTIC

A third,

EDUCATORS
SHIRK
DUTY

You no doubt have gotten the idea and will not be surprised to know that another is

EXPOSE!
MATHE
LIES

All of these are in support of the sign at the top of the post:

CIRCLE
NOW
SQUARED

$$\pi = \frac{9}{5} + \sqrt{\frac{9}{5}}$$

COPYRIGHTED 1967
PROOF
AVAILABLE

That value of π is $3.14164\ldots$, quite close for an approximation using only four one-digit integers, the reason being that the solution of $x+\sqrt{x} = \pi$ is $1.7999648\ldots$, a striking near-coincidence.

C.'s text includes a passage that is worth quoting at length, since it shows, I think, how many cranks' minds work.

Amateur?   Idiot?   Genius?

When these newly found truths of mathematics are considered in a logical manner.

You will find that the author is neither an amateur or an idiot in these matters.

And that the present day mathematicians and their predecessors have erred in their findings. When they have erroneously proven that these truths are not possible. And have emphatically stated that only an amateur or an idiot in these matters would consider otherwise.

This stigma has now been reversed. It is now idiotic for mathematicians to ignore these new truths.

Due to this stigma. A great number of mathematicians throughout the country. Have refused to review these findings. Since they consider it a waste of time and effort to review such absurd findings.

Mathematicians who have reviewed these findings. Have neither affirmed or refuted these findings. Why have they remained silent in this very controversial matter? Since these truths satisfy the rigors of mathematics.

Is it because they do not have enough courage to reveal these truths in an open manner? They should be able to pass judgment in accordance to the rigors of mathematics.

Eventually; these truths will be reviewed by a leading mathematician who will issue a commendatory. And they will become a part of our textbooks.

Although this will cause publishers to correct millions of dictionaries, technical books and school textbooks.

Why did not the mathematicians discover these truths in the past?

Primarily; because they did not approach these solutions in the proper manner.

Secondly; no one dared to even consider that it was possible for great mathematicians to have erred in these matters.

The second edition of C.'s pamphlet does not include that last statement, but it does include something not in the first edition:

Calling a Spade a Spade;

Our educational system is filled with too many inadequate teachers who are only interested in their unearned pay.

Our educational system is filled by too many professors who primarily are only interested in their pet projects.

Perhaps true, but C. probably came to those conclusions because no professor would listen to him.

I once made an error that I will not repeat: I mentioned to an angle trisector, P. M., that it was impossible to construct a square with the same area as a circle using straightedge and compass alone. I did this to illustrate the point that some problems in mathematics cannot be solved. Mentioning circle-squaring was an error, because M. immediately turned his attention to the problem and solved it to his satisfaction. I then made another error: his construction was so crude that putting it on graph paper and counting squares was enough to show that it was not exact, and I told him so. He responded by producing another construction, one that could not be refuted by counting squares. The lesson to be learned is not to mention unsolvable problems to cranks, because they will solve them.

M.'s first construction is contained in a booklet, *Noah's Arc*, hand-illustrated by the author, that is not written in the common quadrature style. It starts:

A few thousand years ago a great rain of mathematical ideas suddenly flooded the earth.

Now in this time some men said that the earth was flat and square. Others argued that it was circular. In that day and age there lived a simple but very wise man. His name was Noah, and he was determined to *reconcile* the square-earth and round-earth problem.

Noah knew intuitively that a flood of mathematical proportions was coming. Before the onset he constructed an *ARC*.

Then he gathered up all the symbols, signs and major segments of the mathematical arts; tied them up with a few chords; and placed them in his *ARC*.

He then set sail upon the rising waters of the Sea of Doubt.

That is clever, and charming. By the time the flood was over Noah and his Arc had squared the circle. The construction is not difficult (Figure 40). Take the circle with two perpendicular diameters and strike two arcs, centers A and B, with radius equal to the diameter of the circle. Their intersection at C, M. said, gives one

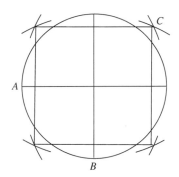

FIGURE 40

of the corners of the square with area equal to the circle. Since this implies that $\pi = 8 - 2\sqrt{7} = 2.708\ldots$, it was not hard to improve on.

His improved constructions gave improved values of π, but M. was told that they were not right either. His reaction to the continued rejection of his values of π was original, though natural when you think of it: he decided that π was unnecessary, and the circle could be measured without it. If you grant that, then of course the exact value of π is merely an unimportant detail. His booklet, *There's Pie in the Sky*, also charmingly hand-illustrated, has a short history of π, even mentioning the first user of the symbol, William Jones. It concludes with formulas for the circumference and area of a circle that do not use the unnecessary π:

$$A = d^2 - r^2 + \frac{4d^2 - 2r^2}{100} + \frac{4d^2}{10,000}$$

and

$$C = \frac{2A}{r}.$$

There: π is back in the sky where it belongs and we have formulas for the circumference and area of a circle that do not involve that pesky and irrational number. However, if you substitute $2r$ for d in the first equation, what you get is

$$A = 4r^2 - r^2 + \frac{16r^2 - 2r^2}{100} + \frac{16r^2}{10,000}$$

or

$$A = 3.1416r^2,$$

which bears a close resemblance to $A = \pi r^2$. You may try to put π in the sky, but it will come back.

The life of a circle-squarer, M.'s in particular, can be hard:

> Received your letter in the mail today and I will try to answer it as calmly as possible, but since I also received one today from [R. N.] of [a scientific] magazine which left me boiling I may find it hard to simmer down to a mood of pleasantness. I asked Mr. [N.] to please send me a simple statement as to what the Ancient Geometers called THE TRADITIONAL RESTRAINTS and he didn't even have the decency to use my stamps for that, but instead sent me a very flippant dissertation on why he couldn't do this and that. Anyway, I've dispensed with his nonsense.

It is necessary for a circle-squarer to be resilient:

> I know that most people think I'm a crackpot, but it flows off my back like water from a duck's. I believe in simplicity and these far-flung ideas of most mathematicians don't belong in my mind. Getting back to my expectations, I had really hoped that some colleges or even high schools would let me come to their establishments and demonstrate my idea, on the blackboard, to them. A face-to-face confrontation would have been an ideal thing, I think. But since I've had time to reflect on the whole thing I'm much better off down here in the country, knowing that I've solved a problem that they say has never been done in over 2,000 years of math.

It is not easy for a non-crank to understand how circle-squarers can get around the fact that circle-squaring is impossible. Most of us cannot hold in our heads simultaneously the truth of both a proposition and its opposite. Two plus two cannot be four *and* five: we must make a choice. However, anyone who has already accomplished one impossibility is not easily going to be stopped by another. The true circle-squarer manages to show, to his satisfaction, that the proof of impossibility does not contradict his construction.

Often the circle-squarer's thoughts on resolving the contradiction are impossible to follow, but once in a while it is possible to get a glimmering of what is going on in his head. There follows an example that shows what I think is a common way of dealing with the fact that π cannot be both $3.1415926535\ldots$ and the different value the circle-squarer has found. It is to deny that π is really a number, or that its numerical value has anything to do with squaring the circle. The reason this circle-squarer, W. L., is easier to follow than most is, I think, that his native language was Spanish, so the wind-filled and meaning-free abstractions that come naturally in

English (so naturally that many writers of English, not all of them cranks, are unable to express themselves without using them) were not available to him. Writing in 1986, he said:

> I am [W. L.], an artist by profession. I am not a mathematician nor a writer, but I have a passion for geometry. [In] 1981 I discovered a geometrical construction for finding the area of a given circle, just using a straight-edge and compass, without using any algebraic formula. It is so elegant, so simple. I was astonished by its accuracy. I spent four years trying to disprove what I had already discovered, for I might be wrong; but could not find a single contradiction to contest its validity.
>
> I know that in mathematics squaring the circle was proclaimed impossible, I have discovered a way of constructing such, and I have the concrete proof of its validity. My only regrets with my discovery is that it will dethrone the already accepted approximation of π. My exact solution to this problem of squaring the circle, disagrees to the point where the true value of π is concerned. It flatly contradicts the present accepted value of π. I am already at the verge of despair and felt like giving up further research since I am scared at the thought of disproving the already accepted value of π by eminent mathematicians.

L. was skilled enough to find that his construction gave π a value of $3.1446\ldots$, one-tenth of one percent greater than $3.14159\ldots$. So, he denied numbers any place in geometry:

> Based on my research, I have found the following:
>
> (1) That I can find the area of a given circle merely by construction; without using any numbers. The construction is by pure logical reasoning, using Euclid's axioms and theorems.
>
> (2) That Euclid's geometry is absolute and numbers have no meaning.
>
> (3) That π's only position in mathematics is its relation to infinite series. That π has no relation to the circle, whatsoever. Its true concept rather is arithmetic, not geometrical.
>
> (4) That all the propositions contained in Archimedes' Book of Lemmas are all intended for squaring the circle; but he was not able to prove it because he was convinced that the only solution in finding the area of a circle was to find a constant, the ratio of the circumference of a circle to its diameter, as he did in his *Measurement of the Circle*, where he computed the value of π between $3\frac{1}{7} - 3\frac{10}{71}$ and he introduces numbers as his tools in geometry. I was lucky indeed for he gave me the chance to discover the required construction.

It is possible to see how L. was thinking, a little. His construction starts with a circle and ends with a square with the same area, proved to be the same, he thought, using only the axioms, postulates, and theorems of Euclidean geometry and not using any numbers. The proof refers only to points, lines, circles, and a triangle or two. No numbers! Numbers have nothing to do with geometry! Euclid said that the sum of the angles of a triangle is two right angles, not that it is 180°. The area of a parallelogram is the product of its length and height; the volume of a right circular cylinder is the area of its base times its height. There are no π's there, and no numbers. I think that if you tried to argue with L. and lead him down the path that leads from geometry to $\pi = 3.14159\ldots$, he would refuse to take the first step. Geometry has no numbers in it, so why, he would ask, are you trying to bring numbers in? It is as if L. had built a bridge over a river and you came to tell him, on philosophical or logical grounds, that his bridge was impossible. Philosophy has nothing to do with building bridges, L. would say; look at it, people are walking over it, and philosophizing cannot change that.

Just as L. was skilled enough to calculate his value of π, he was rational enough to realize that the question of what π really is, if it has nothing to do with squaring the circle, needs to be answered. He gave it a try. His list of findings contained:

(9) That I have discovered the required general construction of squaring the circle, using only two instruments: a straight-edge and a compass by finite Euclidean means, without using π. My method is completely general and can be applied to any circle.

(10) That $3.14159\ldots$ (approx. value of π) has no place in Euclid's geometry, since π is an approximation. π's only domain shall we say is non-Euclidean. Numbers, integers, and π create their own geometry. If we insist that π is correct then we create another geometry different from Euclidean geometry and I have proof for this. It is futile for anybody to find such construction of squaring the circle if we believe in the present value of π, for I have labored so much finding such construction basing my computation on the value of π. I have encountered constants (numbers) that verify the correctness of π but without geometric figures. I have gone thru the ordeal of sleepless nights penetrating its roots and translating my findings thru geometric figures but it seems that my figures are bending and not flat; hence non-Euclidean.

That is an argument that would be hard to answer. L. is saying that when we use π, we depart from Euclid's π-free geometry. Things bend a little, and when things bend, the value of π changes. Think of standing at the North Pole and wanting

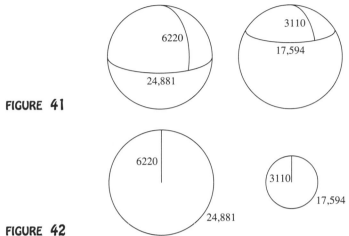

FIGURE 41

FIGURE 42

to calculate the value of π. You could walk down to the equator, keeping track of how far you go, and then walk around the globe at the equator until you came back to where you started. You would have experienced walking from the center of a circle to its boundary and then walking around the boundary. You would have two numbers: the circumference of a circle (the distance around the equator) and its radius (the distance from the North Pole to the equator). (See Figures 41 and 42.) It would have been a hike of 6220 miles to get to the equator and then another 24,881 miles around it, so you would calculate

$$\pi = \frac{24{,}881}{2 \cdot 6220} = 2.00008.$$

If you had hiked only down to latitude 45° north and then circumambulated the globe, you would have walked south for 3110 miles and then around a circle with circumference 17,594 miles, giving a ratio of

$$\pi = \frac{17{,}594}{2 \cdot 3110} = 2.828617363.$$

This is closer to 3.14159 . . . because the bending is less. It is left as an exercise to the reader to show that the value of π gotten by circling the globe at north latitude θ (measured in radians) is

$$\frac{2\pi \cos \theta}{\pi - 2\theta}.$$

It is also left as an exercise in applying L'Hôpital's Rule to show that

$$\lim_{\theta \to \pi/2} \frac{2\pi \cos \theta}{\pi - 2\theta} = \pi.$$

A final exercise for the reader is to calculate how much bending would be needed to change π to L.'s value of $3.1446\ldots$; that is, to find θ so that

$$\frac{2(3.1446) \cos \theta}{3.1446 - 2\theta} = \pi.$$

Consider how you would answer the following, on L.'s terms.

> What I am saying here may sound exaggerated, but it is the truth. When Evariste Galois first introduced some of his ideas, very few would believe him. Gauss never published his non-Euclidean geometry because of his belief of Kantian philosophy. Lindemann proclaimed the squaring of the circle impossible; but Lindemann's proof is misleading for he uses numbers (which are approximate in themselves) in his proof. Mathematicians try to prove this problem numerically, notwithstanding the fact that the true nature of the problem of squaring the circle is geometric. For me Euclidean geometry is absolute, whereas non-Euclidean geometry is only an illusion created with numbers. The present value of π creates such geometry, like the gravitation of Newton and Einstein. In Newton's gravitation the background is there, the actors (body of matters) act on it, while in Einstein gravitation the actors create the background. So π is the actor and it creates the background (non-Euclidean) where numerical manipulations obey to such background, and no one can detect that all are bending and it departs from Euclidean plane geometry which is absolute, therefore π has no place in Euclidean geometry since it is a number.

You could not answer on L.'s terms. It would also be impossible to blast L. out of his frame of reference. Cranks accomplish the impossible, but we cannot. You can deny that π bends space, but can you *prove*, to L.'s satisfaction, that it does not? Until you can, you will make no impression on L.

A Collation of Geometrical Principles and Propositions and a Combination of Geometrical Figures by which the Ratio of the Diameter of the Circle to the Circumference is Found to be as 1 is to 3.25 is the long title of an eight-page pamphlet published by D. B. in 1881. B.'s value of π is one of the largest ever seriously proposed. Such inaccuracy is all the more surprising because B. was no mean mathematician, the first step in his construction being,

> Draw a cycloid and bisect it by the line IC.

Not many people know about cycloids, and there are professional mathematicians who would have to stop and think if asked to draw one.

Here is his reasoning that the customary value of π was in error:

> It has been asserted "that 3.1415926, the approximate area of the circle which is two in diameter, differs from the true area less than any assignable quantity, which means less than .0000001." Taking into consideration the fact that there are thirteen operations which consist in the extraction of the square root, which can only be approximated, and the terms thus found are used as terms of proportion to complete the work, it looks unreasonable to ascribe such a degree of accuracy to it, as there is nothing upon which to base such an assertion; and the demonstration must be left to the future.

B. was perfectly willing to have two values of π coexist:

> Although the ratio here given may be found to be correct, it will perhaps not come into general use except for mechanical purposes, because it spoils so many fine theorems that have become venerable from age and long use. It is published for the pleasure of those who are fond of investigating both the curious and the useful; that it may be tested and its merits or worthlessness determined.

B. seemed to echo W. L.'s idea that numbers are fuzzy: B. seemed to think that they are approximations only and exactness cannot be attained. Thus, 3.14159 and 3.25 are both only approximations of something that cannot be known exactly, and one approximation is as good as another. That idea may come from the difficulty of measuring things to more than two significant digits with rulers. B. and L. may not have grasped that the world of mathematics is not the physical world of fuzzy measurements and inaccurate rulers.

A 17-page manuscript, *Squaring the Circle*, written in 1973, is an example of the common, or garden, geometric quadrature. Its author mentioned the three famous problems and continued:

> Yes, they were thought to be impossible. And what I am trying to do is to prove to you that one of them is a possibility. And as you can see in my title, I chose number (2). It is a possibility that it leads to an answer for all 3 of them.

There followed many pages of description of a construction. Finally, on page 15,

> Of course it is going to be four points. And if you connect them together they will give you the final answer, creating a square that is going to equal if you square it to the area of a given circle.

Quadrature by assertion is all it was, with no indication that the author knew that a proof was needed or, for that matter, that he knew anything about proofs, geometric or otherwise. The construction that took fifteen pages to describe amounted to taking a circle with radius $r = |OA|$, extending OA so that $|AB| = |OA|$, and then bisecting AB twice to find C (Figure 43). Then, making $|OD| = |OE| = |OF| = |OC|$ gives the other vertices of the square. Since $|OC| = \frac{5}{4}r$, the Pythagorean Theorem gives

$$|DC|^2 = \frac{25}{16}r^2 + \frac{25}{16}r^2 = \frac{25}{8}r^2,$$

and that is the area of the square. That is, one of the most common, or garden, values found by circle-squarers.

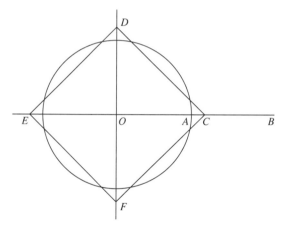

FIGURE 43

[C.'s] System and Process of Squaring the Circle, Together with Methodical Rules for Illustrating General Problems is the title of the 1888 second edition of a pamphlet by A. C., first published in 1868. By assertion, $\pi = 3.125$. There are other assertions as well:

> This system of mensuration is designed by the Author to enlighten the world not only in minor matters, but comprehends and measures that of the universe, curtailing the distances and sizes of the heavenly bodies by trillions of miles and bringing everything down where they belong as measured by the rules of squaring the circle. I find that the diameter of the universe measures by the Square of the Circle only 1,728,000 miles from sky to sky through the diameter of the earth. Consequently the sun,

moon and stars must from necessity be included in this space, which is 13,824,000 times as large in volume as this earth. This actually makes a large world which we live in notwithstanding the curtailing of its artificial dimensions.

C.'s universe is tiny indeed, consistent with cranks' preferences for simple and comprehensible worlds, and anyone with a pocket calculator can find out if his earth is equally tiny. C. said that, if r stands for the radius of the earth, then

$$\frac{4}{3}\pi r^3 \cdot 13{,}824{,}000 = \frac{4}{3}\pi \left(\frac{1{,}728{,}000}{2}\right)^3 .$$

The equation is

$$r^3 \cdot 2^{12} \cdot 3^3 \cdot 5^3 = (2^8 \cdot 3^3 \cdot 5^3)^3,$$

from which it follows that $r = 3600$ miles exactly. Since that is close to the truth and is independent of the value of π, it might have been hard to persuade C. that his universe is too small.

Geometrical Square Root; a Circle Quadratured, and Other Problems, by N. C., is another nineteenth-century quadrature, this one published in 1879. Besides squaring the circle, C. solved eleven other problems that had never before been solved because, C. said,

> Every attempt to solve these problems which has come to my notice, has proved a failure, for the reason that a fact cannot be found by the use of an imperfect theory.

True, and applicable to C.'s work. His quadrature is quite simple and could be used as an exercise in a trigonometry class (Figure 44). Draw the equilateral triangle ABC. If the radius of the circle is 1, then, C. said, the length of AC will be π. Actually, it is

$$\frac{2}{3}(3 + \sqrt{3}) = 3.1547005\ldots,$$

a value that has been found by other circle-squarers.

Square of the Circle, Measurement of the Globe, Decimal Degrees and World-Time is the quadruple-barreled title of a 10-page pamphlet by B. J., printed in Missouri in 1869. One of the unnoticed good effects of television is that people now watch it instead of producing pamphlets squaring the circle.

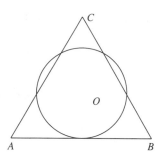

FIGURE **44**

J. said that $\pi = 3.12$ exactly, by assertion, and also that

> The number of the hours of the day should be unchanged, but the division of the hour should be twenty-six (26) instead of sixty (60).

The reason there should be 26 minutes in the hour is that $312 = 12 \times 26$.

Television has not entirely destroyed circle-squaring pamphlets, as shown by π—*A New Value*, by T. J., a 30-page printed pamphlet published in 1981. J. was on the faculty of a Polytechnic in England, in its Department of Civil Engineering, and his pamphlet is full of elaborate constructions. It was not necessary to go through them since J. had the technical skill needed to calculate, to sixteen decimal places, the value of π that his constructions determined, and here it is:

$$3.14358024977942096\ldots.$$

J. maintained that approximating π by calculating circumferences of inscribed and circumscribed polygons was not valid: only a straightedge and compass squaring of the circle would do. He knew a lot of mathematics, and some history, sort of:

> Thus we can conclude that the Pythagoreans and Plato knew how to "square the circle" and that is why they insisted that the circle must be squared by the straightedge and compass alone and that the "process of exhaustion" is not acceptable. . . .

> The emergence of the present value of the circle constant stems probably from the fact that the Pythagoreans, who formed themselves into a secret society, refused to share their knowledge with outsiders. A candidate Pythagorean had to pass the "Pons Asinorum" entrance test and to take the solemn oath to preserve as secret the knowledge accumulated by the Pythagoreans.

J. said that Archimedes also knew the secret Pythagorean value, that trigonometric and other tables would have to be corrected, and that it might be possible to trisect the angle.

Most circle-squarers find one value for π and defend it, if not to the death, at least against all comers. J. V., however, was more flexible. In October 1982 he wrote:

> The Quadrature of a Circle, never done before since mathematics became a true science to man in this world, is going to bring you knowledge you never expected in your life time. . . .

> With the "Quadrature of a Circle" I have come into a new essential in mathematics, which I now call it:

> CIRCLENOMETRY

> . . . In my UNIT CIRCLE, Radius = ONE. "PI" = the true value of 2.91419041.

Seldom does one find π to be less than 3.

By November 1, 1982, V. had changed his mind:

> In geometrical form, and with exact symmetry, today I have "Squared the Circle", the classical problem of all times, and consequently found that the proportion of the circumference of a circle to the diameter is EXACTLY 3.0625. All the books in the world that show 3.1416 as the quantity are definitely wrong.

On November 28, 1982 there was another revision:

> CONFIRMING my work on the *ONE* "Quadrature of a Circle",

> $$\pi = 2.96000000000008$$

> I have checked, rechecked, this value very closely, and its figures, being the circle the "Perfect Form" in "Plane Geometry", and a truth from a

> *Greater Intelligence.*

December 3, 1982, less than a week later:

> Being the ONE "Quadrature of a Circle" the highest mathematical achievement of mankind, ever, and I, myself, with a profound sense of its scientific importance, for once and for all, in a definite final way do attest the "Quadrature of a Circle" as a true FACT.

> The first man ever to work the area of a circle, mathematically, *without* the value of "PI", I do further attest that the true *one value* of the proportion of the circumference of a circle to its diameter is, "PI", that is, the total value of 2.91421351481511+, no buts, no more than a little less, period.

So V. had come full circle, from 2.914 back to 2.914.

You might think that definite final way, period, would be the end of it. Not so. April, 1983:

> With all due respect, if I may beg your attention for a few seconds, you must be one of the first to be informed that I have today a definite "geometrical proof" and *numerical evaluation* of the fact that the "mathematical constant" of the Universe, nature, the sphere and the circle is for *true* the "*square root*" of three (3), that is, 1.7320508+.
>
> That means, that in any circle, the area of the circle is $3R$, and in any sphere, the volume of the sphere is $3\sqrt{3}R$, where R is the value of the radius. That is, that in this Universe where nature's only "show" is spherical, and circular movements, this *mathematical constant* of $\sqrt{3}$ is in complete control of all geometrical calculations in unfolding the mysteries of the symmetries of such.

That is, if $R = 1$, then $\pi = 3$ for the circle and $\frac{9}{4}\sqrt{3} = 3.89711\ldots$ for the sphere. Of course, if $R = 2$, we get different values entirely; but let us be charitable and suppose that for the circle "R squared" was meant, and "R cubed" for the sphere. Even so, four (or five) different values for π is a world record, especially when found in less than six months.

H. W. of Ontario squared the circle on three sheets, individually notarized, in 1981. By assertion,

$$\pi = \sqrt{17} - 1 = 3.123105626\ldots.$$

The sheet on which this appears has a box containing

> This presentation should
> *not* be considered to be
> the geometrical proof.

which it certainly is not. I have not seen the proof.

In 1931, E. Y. published *The True Value of Pi: The Circle Squared*. He wrote in capitals for emphasis:

> NOW FIRST ACCOMPLISHED AND PROVEN BY THE PROCESS OF PURE ARITHMETIC, AFTER A FULL THREE THOUSAND YEARS (SINCE WHEN THE ANCIENT MAGI KNEW AND USED IT IN THEIR LODGES), ALTHOUGH LEGIONS HAVE TRIED FOR IT AND FAILED, LIBRARIES HAVE BEEN WRITTEN ABOUT IT PRO AND CON, AND SCIENTIFIC GOVERNMENTAL REPORTS HAVE EMPHATICALLY STATED THAT IT COULD NOT BE DONE.

Immediately after that came:

ASSUMPTIONS:—
That the value of π is arbitrarily fixed at 3.14175625 as correct.

Never has there been a bolder, or balder, quadrature by assertion.

π *ist rational*, by O. Z., a Modellbaumeister, was published in Germany on 1983. Z. was another of the rare circle-squarers who changed his view on π over time. Before September 1975, he inclined to 3.1415926535576; from then until January, 1976 it was 3.141592653598; but since then it has been 3.1428 exactly. It is important, he said, that this is $97 \times .0324$.

These reports of circle-squarings at second hand are efficient, sparing the reader the labor of extracting what may be a very small essence from what may be a very large volume of text. But they cannot replace the experience of confronting an actual quadrature, raw and untamed. Anyone who can get to a college library can have that experience because I think that C. H. distributed copies of his 128-page hardbound book

BEHOLD!
THE GRAND PROBLEM
NO LONGER UNSOLVED

subtitled

THE
CIRCLE
SQUARED
BEYOND
REFUTATION

to every college library in the country when he published it in 1931. H. was unusual because the quadrature ($\pi = 3\frac{13}{81}$) was not his; H. was a disciple of C. F., whose work he was publicizing. The book is considerably more impressive than the usual quadrature, but it is like many of them and is well worth looking at if you want to look at such things.

R. W. was not happy with numbers:

> Many years ago I became quite displeased with decimals. Why? Because they had a sloppy, unending way of trying to relate one thing to another. They are all right in their place, as long as they come to a definite point, but what are we going to do with those infinite 3's and 6's and the like? These many years back that I mentioned found me constantly replacing infinite decimals with either a straight line, or an arc, or a combination thereof, or a ratio.

Suppose we have two straight lines of known length. One is one inch long and the other is three inches long. In turning them into numbers we find that we express them either as 1:3 or as .33333333333333333333333333.

Which is better?
HOW IMPORTANT IS 3.14159ETC?

W. was one of the frustrating cranks who knew the answers but would not tell them:

The number 1 appears to be a good solid number. Appears to be. But what kind of a number is it that can be squared and/or cubed and give itself as the correct answer? To see 1 as it really is maybe we should put it in ratio.

1:2 as 4:5 as 62.5:64 as 125:128 as 1:1.28 as 1:1.024
What nonsense is this? Is it right or wrong?
I'll never tell.
If correct, what does it mean?
I'll never tell.

There is a hint that $\pi = 3.256$:

In giving it some thought we might wonder if Detroit is squaring with us when it tells us that it rates the speedometers a little fast in order to keep people from going as fast as they think they are. Are they telling the truth, or are they merely designing speedometers for cars that should use 8.00 tires and end up with 7.60's on them? Or is it possible that these engineers—that have graduated from your colleges—haven't heard of 3.256PERIOD? Does that figure really mean something? Time might tell, but I never will.

W. had also duplicated the cube and trisected the angle, and was knocked slightly off center when he got a letter about the trisection:

[I] have just received a report that the trisection of an angle is impossible!

The Massachusetts Institute of Technology has duly informed me—(via a medium grade paper bearing a poorly centered mimeograph message)—that there is "rigorous proof of the impossibility of a solution for the trisecting of an angle".

To regain his equilibrium, he fell back on the legend of the Patent Office official who decided that everything had already been invented, and on

How can a person honestly tell of something he honestly knows nothing about?

He added:

> Suppose a person earns his living wage teaching mathematics. He receives his pay for explaining true situations to other people in such a manner that they will grasp and hold true these same situations. Thus he earns an "honest" living wage. This person, in his work, however, must be solid enough in conveying his truths only to the point where he is malleable in accepting stronger truth. Now since mathematics is a rather "definite" science—to say the least—it would appear that anything in that science that is "definitely unsolved" would have a wide open door ready for any possible solution. However, if we close the door to all possible solutions, we are clearly showing that we are no longer malleable. And, since we are being paid for our malleability, we are then at that point obtaining money under false pretenses. Naturally, if we check the solution and find it untrue, we maintain the thing for which we are being paid.

Cranks can be knocked off center, but they usually recover. M.I.T. was wise to devote no resources beyond a poorly-centered mimeographed message to him.

Teachers of calculus are all aware that the idea of limit, so simple and clear to them, nevertheless reduces many beginning students to dumbness, confusion, or incoherence. *The Quadrature of the Circle Perfected*, by C. G. (New York, 1868), shows that things were no different in the old days:

> Sir Isaac Newton, Lagrange, La Place, Ferguson, Kepler and others might have been relieved of those tedious struggles with the question, if, instead of seeking a solution of it by the Differential Calculus and in the employment of infinitesimals, inscribed and circumscribed polygons, &c., they had taken a standpoint outside the circle, and surveyed it through the geometrical media afforded in these diagrams. They did, indeed, perform wonders in the methods they so ingeniously devised; and they merit vast praise and gratitude for having built up a system of mathematics, which has so largely contributed to the progress of other sciences and the useful arts. It was a terrible leap, however, which Sir Isaac took, when, to relieve himself of the constantly recurring difficulties which beset his efforts to measure the length of a curved line, by applying directly to it straight lines, he asserted the strange untruth that the ultimate ratio of an enlarged circle to a straight line is the ratio of equality.

Ferguson? Who was Ferguson? Whoever he was, he did not belong in the company G. put him in.

G. was quite explicit about his value of π: it is

$$\frac{10\sqrt{2} - 11}{4},$$

or $3.14214\ldots$, off by a mere $.02\%$.

When W. T. squared the circle in 1986, he enclosed with his 5-page quadrature a color photograph. It is of two television sets, one on top of another, each showing a black and white picture of a smiling actor holding up a can of Pledge. The picture is enclosed by a curved border that was elliptical on the top set and more or less circular on the bottom set. T.'s caption is

Photo shows TV adjusted for $\pi = \sqrt{10}$.

This was undoubtedly the first time π had ever been linked to television reception. T. may have meant the picture with the circular border to be the one that was using his value of π, but I think that the elliptical one was more faithful to the signal that the television station sent out. The actor's face in the circular picture is too long, and the can of Pledge too skinny. However, the circular picture is considerably clearer than the elliptical one.

T.'s quadrature is by assertion. You might wonder how it is possible to fill up five pages with an assertion that π has a specific value. The following excerpt from a letter sent to him by a patient mathematician shows what proofs by assertion sometime involve:

I daresay you will not receive many replies if you have sent this letter to mathematics departments. Mathematicians have learned—in some cases by bitter experience—not to get involved in discussions with eager amateurs who do not understand what constitutes a mathematical proof. Perhaps like a fool, rushing in where angels fear to tread, I shall attempt to show you your error. Your letter asks us to try to find "3 new solutions of π." Frankly, sir, this doesn't make sense. One does not "solve" numbers. You can only solve equations. From the three following pages I infer that you believe that you have found three "solutions of π." May I take the liberty of clarifying your intent? Is it not the case that you really mean you have found three proofs that π is $\sqrt{10}$? I take this to be your meaning, and will now attempt to show you why you do not have any such proof. Since the error is the same in all three cases, I will limit my discussion to your first argument. Starting below the dividing line on page 317, you have first of all a correct statement of the relations among the various figures you have drawn. Your second equation, in which the symbol π appears twice, is also a correct statement of the areas of these figures.

At that point you say, "Assuming $\pi = \sqrt{10} = 3.162278$" and proceed to replace the symbol by that value in *one* of the places. You then solve the equation correctly for the *other* occurrence of π and find—to no one's surprise—that the resulting value is the one you assumed in the first place. I should tell you that this is not the first time I have read such an argument. What is a continual source of wonder to me is the following: Why do you find it convincing that *assuming* a value of π allows you to obtain that value? Surely you must realize that you have assumed what you claim to be proving. In fact the original equation, in which π occurred twice, is an identity. It will be a true equation *no matter what value you assume for π*. I never cease to marvel that no one who uses this argument ever notices that if you assumed $\pi = 47$ or any other value whatsoever in one of the two places, the value assumed would result when the equation is solved for the other occurrence.

The patient mathematician went on to explain that Archimedes had shown that π was smaller than $\sqrt{10}$, and that Lindemann had shown that π was transcendental and thus could so not be equal to $\sqrt{10}$. Toward the end of his long letter, his patience began to wear thin:

> Your extraordinary (completely ungrammatical) claim on page 320 that "π should be redefined as the area of a circle of unit radius of the base circle of a pendulum" comes as close to being meaningless as it is possible to come. Please forgive me if I speak with absolute candor. The twittering of sparrows and the writings of great scientists have one thing in common: both are incomprehensible to the average person. One should not conclude, however, that the sparrows are too profound for human beings to understand their thoughts. They just don't have any thoughts in the ordinary human sense of the term. I regret to inform you that these few sentences of yours are of precisely that type. One cannot simply sling words around without regard to their meaning.

The mathematician's letter probably did not do much good, other than the good it does to practice one's prose, since T., who billed himself as "architect," also included portions of his plans for a Peace Memorial, a building whose base was to be one mile in diameter, in the shape of a pyramid 2000 feet on a side. Some of T.'s sheets contained comments like:

> Car traveling across diagonal of time plane must increase speed 41.4% to adjust for misalignment on earth. Immediate stoppage of work involving atomics & heavy magnetics.

Others had seemingly random and irrelevant comments scattered across them. T., while probably not certifiably insane, was very likely one of those people who cannot be reached by reason.

On the other hand, he may have had a point when he wrote:

The 5 Platonic Solids are too perfect for π to be 3.141592.

The method of proof used by T.—namely taking an identity involving π and showing that substituting 3.125, 22/7, or some other value into it produces a true equation— is not rare. M. K., a resident of Pakistan, produced a small pamphlet in 1985 titled *An Amazing Discovery*. He sent it to the National Aeronautics and Space Administration, and you would never guess to which subdivision of NASA it was referred. It went to the Director of the Office of Small and Disadvantaged Business Utilization, who dutifully replied, in part,

Your analysis is very interesting, but at the moment has no applicability to NASA programs.

K. said that π was 3.125 because that was the value he had discovered, and he went on to show why it was right:

Let me have to prove that the above mentioned fantastic value is *exact* and gives constant consequences.

Suppose, there is given a circle whose circumference is exact 4 units. Find out its diameter with the help of said value:

$$d = \frac{c}{\pi}, \qquad d = 4 \div 3.125, \qquad d = 1.28 \text{ units.}$$

Now, find out the circumference of a circle whose diameter is = 1.28 units with the help of said value:

$$c = d \times \pi, \qquad c = 1.28 \times 3.125, \qquad c = 4.$$

Difference of circumference = 4 − 4 = 0.

I, therefore, have firm conviction that the discovered value of π = 25/8 = 3.125 is exact.

I am sure that it will gladly be accepted on the basis of reality of fact.

Thanks God! I have the honour to produce such a value of π for which world's Mathematicians had been dreaming since thousands years.

It is always easy to deal with 3.125 circle-squarers: refer them to Archimedes' proof that π is greater than $3\frac{10}{71}$ and hope that will take care of them. K. acknowledged receipt of the copy of Archimedes' argument that I had sent him, and ended his letter with:

I shall be grateful once more, if I may be kindly be informed the relation among both subjects i.e. Geometry and Astrology.

My answer that there was none may have made him seek greener pastures elsewhere.

Mathematicians all know that there is no big money in mathematics, except as recompense for producing a best-selling calculus textbook. Thus this 1988 letter from a lawyer is highly unusual:

A LEGAL AND BINDING DOCUMENT
RELATING TO THE TERMS AND CONDITIONS SET BY
[R. R.]
CONCERNING THE DONATION OF
FIVE HUNDRED THOUSAND DOLLARS
TO A UNIVERSITY, COLLEGE OR SCHOOL
WIIICII CONFORMS TO THE
TERMS AND CONDITIONS OF THIS DOCUMENT

Wow! What do I have to do to get it? Number 1 of the conditions and terms was:

Your establishment is One of TWENTY-TWO UNIVERSITIES, located in the UNITED STATES OF AMERICA, (List attached as Document Six), which has the opportunity of receiving a DONATION, on terms and conditions set, of $500,000, (FIVE HUNDRED THOUSAND DOLLARS), provided, that by the 4th day of JULY, 1988, (NINETEEN HUNDRED AND EIGHTY EIGHT), YOUR ESTABLISHMENT, is the first ESTABLISHMENT out of the TWENTY TWO SELECTED which can produce mathematical or geometrical evidence to disprove the mathematical-logic, geometry, or the postulations contained in the attached five paged SYLLOGISM FRONTISPIECE, (identified as DOCUMENTS ONE, TWO, THREE, FOUR and FIVE), of the UNPUBLISHED MANU-SCRIPT, entitled, "Compatibility of the Linear and Curved Dimensions.", written and compiled by [R. R.].

Disproving evidence of a frivolous nature contending variations beyond the fifth decimal point will NOT BE ACCEPTABLE.

Unfortunately, the $500,000 would not go to the lucky mathematician who refuted R.'s quadrature (for that is what lies behind "compatibility of the linear and curved dimensions"), but to create a scholarship fund

for the ADVANCED EDUCATION of UNDER-PRIVILEGED PER-SONS POSSESSING UNIQUE ACADEMIC GIFTS.

Someone should have gotten the money, since R. squared the circle by assertion. It was not hard at all to look at his four sheets and see that what he was saying was that

$$\pi = \frac{7}{4} + \sqrt{2} = 3.16421\ldots .$$

The only reason R. gave for the correctness of that value was, essentially, "because." It is not even close.

R., or his lawyer, may have escaped having to disgorge the half-million because none of the twenty-two establishments who received his challenge may have been able to refute to his satisfaction the contents of the book of which his four pages was merely the introduction:

> The overwhelming evidence postulated in Premise One and Premise Two and supported by fifteen unique axioms and one corollary together with thirty five sets of unique geometrical diagrams, confirmed by precise mathematics compiled in the following thesis to be published as a book, entitled "Compatibility of Linear and Curved Dimensions."—contains two methods of assessing the area of a circle inscribed in a square by mensuration: Never thought to be conceivably possible.

It is hard enough to convince a crank that he is wrong when he is alone; imagine what would happen if he had a lawyer along! $500,000 would hardly be enough to repay the effort needed.

Another 1988 quadrature by assertion, by D. G., represents for the forces of reason another step backward or, if not directly backward, at least with a significant rearward component. The reason is that G., a retired teacher of high- school English, called in the newspapers, and more than one responded. If you ever pass through Durham, North Carolina, you can stop off at the offices of the *Herald* and inspect the almost full-page spread that it gave G. in its June 6, 1989 issue. The headline is

> Squaring the Circle,

the subhead is

> Schoolteacher Thinks He Has Solved Age-Old Mathematical Problem,

and the caption of the picture of G., posing in front of a large diagram of circles, arcs, and lines, is a quote from him:

> Someone who has figured pi by former methods has taken liberties with significant decimal places along the way.

There is danger that the typical newspaper reader will take away from that the impression that there is doubt and controversy about the proper value of π, and that mathematicians are not sure about what they are talking about. Thus is unreason encouraged, as if it needed any encouragement, and thus does it flourish.

The press, of course, is objective and does not endorse G.'s value for π, which, by the way, is

$$\sqrt{2} + \sqrt{3} = 3.14626\ldots.$$

However, about 95% of the story's space is given over to G., his accomplishments, and his quadrature. G. disposed of the fact that π is transcendental in the manner usual to circle-squarers:

> In 1882, Ferdinand von Lindemann, a German mathematician, proved that the circle could not be squared, he said.
>
> Experts still say that it cannot be done, but they are using calculus, [G.] said. "With this procedure, you can't square the circle," he said, but with geometry, it can be done.
>
> Pi is a geometric ratio, not a number, he said, and must be dealt with in a geometric way, with a compass and straightedge.

That G. was brushed off by the people he sent his work to is also treated "objectively":

> "Science is supposed to be open-minded, not a closed book," [G.] said. "The remarkable thing about all the responses I have received is that none of them address my arguments. They show that it's wrong by using their theories."
>
> Isaac Asimov, a well-known author and scientist, wrote back a curt message saying that the theory is incorrect. [G.] seemed to find that especially frustrating, because Asimov made an obvious error in calculation and based his reason for rejection on that calculation.

The newspaper asked someone who knew something about mathematics—a physicist—about G.'s work, but even so G. got the last word. The story ended:

> Dr. Robert W. Brehme, a professor of physics and astronomy at Wake Forest University, said that [G.]'s argument is incorrect.
>
> "He is as wrong as the fact that 2 plus 2 is 5 is wrong."
>
> Brehme demonstrated with mathematical proof. [G.]'s response to that proof?
>
> "I say measure a circle, if you really believe that. Put it to the experimental test. . . .

"What I'm telling you is that the present value of pi won't stand the test of experiment.

"The circle can indeed be squared."

One reason quadrature by assertion is so popular is that when circle-squarers give proofs, errors in them can be pointed out. If a circle-squarer does not provide a proof, then no mistakes can be found in it. Thus time is saved for other activities. G. P., very active in the 1950s, squared the circle by assertion and had a whole list of mathematical publications, including

The True Value of π and the Fallacy of Archimedes
The Self-Contradiction of the Non-Euclidean Geometry
How to Make the Calculus Intelligible
How Classical Mathematics Deceives You
The Treatment of Parallel Lines Without Euclid's Fifth Axiom
The Demonstration of Euclid's Fifth Axiom
Mathematics and Logic

P. had determined that π is $3.1547006\ldots$, and that led him to discover that Archimedes had shown many centuries ago that

$$3\tfrac{10}{71} < \pi < 3\tfrac{1}{7}.$$

Since $3\tfrac{1}{7} = 3.142857\ldots$, P.'s value was larger than Archimedes' upper bound. P. did not try to evade the issue: since he was right, something had to be wrong with Archimedes' proof.

> In a previous publication, *The Operational Method of Mathematics*, I have determined the true value of π. An extract of that determination is given at the end of this pamphlet, to make complete the presentation of the question. The true value of π is found to be 3.1547006, showing that the classical value of π is false from the second decimal place.
>
> I thought I had done all that was necessary to settle the question of π. My evaluation is short, clear, simple and free from arbitrary assumptions. It is therefore easy to see if I have made any error, and if none can be found, the classical evaluation must be discarded. For the classical evaluation is neither short, nor clear, nor simple, nor free from arbitrary assumptions; and so many defects must hide some fallacy leading to an error in the result.
>
> At least this is what I candidly thought. But it did not turn out that way. And now I find that there is in some mathematical circles an opposition to my value of π, with a well defined policy that could be expressed in this way: "We are satisfied with the value of π, being between $3\tfrac{10}{71}$

and $3\frac{1}{7}$ is 3.14+." This attitude is very clever. It tells me: "You have not proved that your value of π is right, unless you can prove first that the procedure of Archimedes is wrong." They are well confident that nobody can prove that Archimedes is wrong: if he were wrong it would have been seen without waiting more than 2,000 years!

How true!

However, there is a great danger in that attitude, for, if it could be proved that Archimedes is wrong, it would not only prove that his conclusions on π are wrong, but it would also prove that the whole of classical mathematics, which could be deceived for centuries and centuries without knowing the truth, *is not a true science!*

And, unfortunately for classical mathematics, it just happens that I can prove that Archimedes is wrong.

Here is how P. overthrew Archimedes. Archimedes got his bounds of π by taking a circle, inscribing and circumscribing regular polygons, and finding their perimeters. So, P. took a sector of the circle including one side of the n-sided inscribed and circumscribed polygons—AI and $A'B'$ in Figure 45—and said:

Now, in the triangle AIA' we have:

$$|A'I| \leq |AI| + |AA'|$$

then

$$|A'I| < \text{arc } AI + |A'A|$$

and

$$2|A'I| < 2 \text{ arc } AI + 2|AA'|$$

or

$$|A'B'| < \text{arc } AB + 2|AA'|.$$

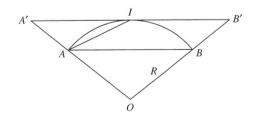

FIGURE 45

But when n is infinite, $|AA'|$ is zero. Therefore, when n is infinite,

$$|A'B'| < \text{arc } AB.$$

This, P. said, is manifestly impossible, since it says that the perimeter of the circumscribed polygon is shorter than the circumference of the circle it is circumscribed around. Thus, P. said, all that can be concluded from Archimedes' argument is that

$$\pi > 3\frac{10}{71} = 3.1408\ldots,$$

and that is perfectly compatible with P.'s true value of π. P. was not awake in his calculus class the day it was demonstrated (or stated) that from $a_n < b_n$ it follows only that

$$\lim_{n\to\infty} a_n \leq \lim_{n\to\infty} b_n,$$

and that strict inequality need not hold in the limit.

On the other hand, P. may have been awake and forgotten. He did not have a high opinion of the quality of mathematics instruction in institutions of higher education:

> In many Universities and Colleges, it is observed that the students avoid, as far as possible, taking courses in mathematics, unless it is compulsory for the kind of diploma they aim to obtain. It is also noted by many professors, that, at a time when there is a great demand for engineers of all kinds, the courses of engineering have a scarce following. Even students taking a course in mathematics may soon get discouraged by their difficulty in dealing with arbitrary and imaginary notions such as the negative number or the limit, which they vainly try to understand and always find unintelligible.
>
> Those who are independent enough to change their plans will switch to another branch of study. Those who are not independent enough to do it, accept with resignation to follow the necessary courses of mathematics, in a spirit of sacrifice to their future. But as soon as they have got their diplomas, they throw away their books of mathematics and they gradually forget all what the college had taught them. Their knowledge of calculus is gone in a few weeks and that of algebra and geometry, if it sinks more slowly, does not disappear less effectively.

P.'s knowledge had definitely sunk.

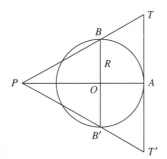

FIGURE 46

After Archimedes had been taken care of, P. could demonstrate why his value of π was correct. Given a circle (Figure 46)

> Draw the tangent at A and unfold arc AB on AT; and draw arc AB' on AT'.

We can overlook the impossibility of doing that unfolding with straightedge and compass: P. was not concerned with Euclidean constructions, he was after the proper value of π.

> Then, if π = (length of the circle)/(length of the diameter) we have:

$$\frac{\pi}{2} = \frac{TT'}{BB'}.$$

Our determination is simply that of a ratio of segments. . . .

> BB' and TT' being parallel, draw TB and $T'B'$ which meet at P on AO. Then PBB' and PTT' are similar triangles and we have

$$\frac{\pi}{2} = \frac{TT'}{BB'} = \frac{\pi R}{2R} = \frac{PT}{PB}.$$

Now $\dfrac{PT}{PB} = \dfrac{\pi R}{2R}$ is satisfied for

$$PT = \pi R \qquad \text{and} \qquad PB = 2R.$$

But then $PB = 2OB$. Therefore, in the triangle PBO we have

$$\text{angle } PBO = 2 \times \text{angle } BPO$$

that is

$$\text{angle } BPO = 60°$$

and then PTT' is equilateral.

From that it is the work of but an instant or two to find that

$$\pi = 2\left(1 + \frac{1}{\sqrt{3}}\right) = 3.1547006\ldots.$$

It does not take too keen an eye to see that the key step in P.'s proof is his assertion that because two fractions are equal, it follows that they have the same numerators and denominators. You might think it would be hard for P. to get around that error when it was pointed out. It wasn't. I quote at length to show how he did it because the passage shows very well the workings of a crank's mind.

Prof. X is an "outstanding" mathematician and the author of a learned textbook on algebra from which he draws both prestige and money. No use to say that he will fight by any means the Reformation of Mathematics which would reveal so many fallacies and errors in his book that it would make it immediately obsolete. So, when the president of his University asked him to answer me, about my "determination of π," he sent me the following letter:

"At the request of President C. . . I make the following comment on your determination of π. I note the following error: from the fact that the equation of your problem is the equality of two fractions, you draw the conclusion that the numerators and denominators of both are identical and thus obtain the identity $PB = 2R$. Everyone knows that two fractions may be equal without having identical numerators and denominators. The fact is (as you could check by making the construction with really accurate instruments), that PB is not equal to $2R$."

This letter of Prof. X, of which a copy has to be sent to his president, is entirely written for the president, to deceive him and make him believe that my determination has an error, without giving any proof of that error which does not exist.

In order that the president, who is not a mathematician, can read the letter, it is written without the least technicality, and without any more symbols than a simple equality.

In order that he may believe that the error exists, he is made to understand that I am an ignoramus, who does not even know what "everybody knows," and that the equation of my problem having several solutions, besides $PB = 2R$ the solution of the problem is not $PB = 2R$, but some other solution; and finally the president is told that a scientific check "with really accurate instruments," *could* show that PB is not $2R$. But the check is not produced, nor even explained, because it is impossible. For, to make the construction *exactly*, it is necessary to know the exact value of π, which is the object of the demonstration; and thus, to contest my value of π, it would be necessary to *assume* that my value of π is erroneous, a fallacy known in logic as the vicious circle.

It would be difficult to write more dishonestly about a scientific matter, and the letter of Prof. X will remain as a monument to the shame of Classical Mathematics. This is why I am publishing it here.

In another place, P. argued that treating his fraction the way he did was correct because, he said, if

$$\frac{PT}{PB} = \frac{\pi}{2},$$

it had to be the case that PT is an integer multiple of π and that PB is the same integer multiple of 2. If the multiple is greater than 2, a value of π that is too small will result. Thus $PB = 2$.

If P. had said that

$$\pi = 2 \left(1 + \frac{1}{\sqrt{3}} \right)$$

just because he said so, he could have saved himself a lot of trouble and gotten exactly as far. The most efficient way to square the circle is by assertion.

There is no conclusion to this section. There can be none, since the end of circle-squaring will come only as a consequence of the end of civilization.

SET THEORY

Crankery can strike anywhere. H. N. was a full professor of mathematics at a research university (though not one of the first rank) when he wrote in 1970:

> Recently I have discovered a surprising fact. Logicians' axiomatic set theory is meaningless in mathematics. They have been working in wrong mathematics and exerting their efforts in vain over the past sixty years. They are still making the same mistake. So I dare say that all the books related to axiomatic set theory must be removed immediately. Only my book describes correct set theory.

His letter was sent to seventeen mathematicians of the utmost eminence and, since he was a colleague, several responded. One polite reply suggested that N. examine works by Gödel, Cohen, Zermelo, von Neumann, Bernays, and Rosser:

> Even if you have already looked at these books, may I suggest that you could again examine them with great profit.

N.'s answer to that was:

> All these books are wrong mathematics.

His answer to another question,

> May I explain why this might mystify many people?

was:

> I know many people who understand me.

His rejoinder to

> Your conclusion about set theory is unclear because there is to my knowledge no one "set theory". Perhaps you mean "There is no axiomatic set theory equivalent to [N.] set theory".

was:

> [N.]'s set theory is the only one correct set theory.

The suggestion

> I urge that you first read the literature [and] then talk with other set theorists.

was rejected:

> I don't have to read wrong mathematics. I have sent preprints to many set theorists, but no one has given me comments, as you have not.

His mind was so closed as to be unopenable. N.'s name disappeared from the membership list of the American Mathematical Society in 1974, perhaps because of disgust with the establishment, perhaps because of death.

SIGNS, THE RULE OF

When we are told, at an early age, that a negative number multiplied by a positive number is negative and the product of two negative numbers is positive, most of us accept the voice of authority and carry on. Some of us even understand when teachers tell us to think of negative quantities as debts, so that if someone gives us four debts of $3,000, we are $12,000 worse off, and that is why $4(-3000) = -12000$. Such explanations can pass over the head of some students, but nevertheless everyone grows up to apply the rule of signs in the correct way whenever it is necessary.

D. I., however, refused to knuckle under, either to authority or to conventional wisdom. He wanted negative numbers to have square roots, but since the rule of signs says that no negative number can be a perfect square, something drastic would have to be done. In *A Challenging of Traditional Mathematics*, a 351-page book (photoduplicated from double-spaced typescript) published in 1981, he decided to change the rule.

> Long ago mathematicians adopted a rule of signs for multiplication that made it a contradiction to think of taking the square root of a negative number. They couldn't figure out what the trouble was, so they made a mystery of the contradiction. No inconsistency of religious dogma has ever been expounded as sacred mystery with greater facility and tenacity than this figment of mathematical supposition. More than that, a vast Cloud-Cuckoo-Land of theory about complex numbers has been built up on a blind acceptance of the meaningfulness of the expression $\sqrt{-1}$.

He pointed out that what he calls the "traditional law of signs" is what results from deciding that it is important to preserve the distributive property of multiplication over addition,

$$a(b + c) = ab + ac,$$

when a, b, or c may be negative. Since $4(7 - 5) = 4(2) = 8$, if we are going to have $4(7 - 5) = 4(7) + 4(-5)$ then

$$8 = 4(7 - 5) = 4(7) + 4(-5) = 28 + 4(-5),$$

so we have to agree that $4(-5) = -20$. If we do the same calculation with 4 with replaced with -4, it follows that if we want to keep the distributive property, then we must have $(-4)(-5) = 20$.

For most of us, that would be that, but I. boldly questioned the importance of the distributive property. He decided that he could do without it:

> The retention of these laws is held to be "justified" on the ground that their retention permits the introduction of negative numbers without abandoning any of the laws that have been accepted as governing operations with natural numbers. In effect, retention of the Commutative and Distributive Laws is justified as an end in itself. There apparently has been little or no interest in the possibility of introducing negative numbers without insisting on retention of all laws governing operations with natural numbers. The argument or implication occasionally encountered that preservation of these laws best serves the convenience of mathematics therefore rings a bit hollow. One may suspect that these laws have been preserved by an act of faith on the part of a discipline that prides itself on accepting nothing on faith.

Students of mathematics accept a great deal on faith, including the necessity of the distributive property. However, with maturity and on reflection, they can see that their faith is not misplaced.

In I.'s system, the product of two numbers would always have the sign of the first number:

$$(4)(3) = 12, \qquad (-4)(3) = -12,$$
$$(4)(-3) = 12, \qquad (-4)(-3) = -12.$$

This has the advantage of making imaginary numbers unnecessary: since $(-1)(-1) = -1$, it follows that -1 has a square root, namely -1.

Of course, we cannot get something for nothing, so we will have to give up commutativity, since $(3)(-4) = 12$ while $(-4)(3) = -12$. And it will no longer be true that equals divided by equals are equal:

$$(3)(4) = (3)(-4) = 12,$$

but the 3s may not be canceled.

There are other difficulties, some of which I. dealt with. The extremely well-known identity

$$(x + y)^2 = x^2 + 2xy + y^2$$

is a consequence of the distributive property, so if the distributive property goes, so does the identity. What it gets replaced with, I. said, is

$$(x + y)^2 = x^2 + 2xy + y^2$$

if x and y have the same sign,

$$(x + y)^2 = x^2 - 2xy - y^2$$

if x and y have different signs and $x > y$, and

$$(x + y)^2 = -x^2 + 2xy + y^2$$

if x and y have different signs and $x < y$. As I. explained, if you take $-2xy$ to mean $(-2)xy$, then you run into trouble since $-2xy$ could then never be positive. What $-2xy$ should mean, therefore, is $(-x)(y)(2)$.

All of algebra will have to be rewritten. I. made a start, deriving what he called the Forward Natural Distributive Law and the Backward Natural Distributive Law, but much work remains to be done. It would make a fascinating project. All quadratic equations have real roots, so the quadratic formula may have to be revised. The fundamental theorem of algebra may be a casualty of the revolution. There may not be much of standard algebra that can be salvaged in I.-algebra, but it would be fun to find out. I. deserves admiration for having the idea of tossing out distributivity and looking at its consequences. It is not an idea that would occur to many people.

I.'s purpose, though, was not to provide harmless and interesting mathematical opportunities. He wrote:

> I am not a mathematician. I have therefore had to ask myself more than once whether I should risk adding my name to the long, sad list of misguided amateurs who have thought to give lessons to master mathematicians. Inexorably, however, my concern with abstract classification theory has taken me into symbolic logic, set theory, and topology.

"Long, sad list" it is indeed.

> For years and years I tried to make sense out of the topology of point sets and couldn't. Gradually, the conclusion became inescapable that something was wrong either with me or the topology of point sets.

Since the topology of point sets seems to be intelligible enough to mathematicians, the simpler and more plausible explanation would be that I am mathematically obtuse. If I publicly espoused the other position, someone might point out that there was this simpler and more plausible explanation. Furthermore, my espousal of the other possibility might be attributed to a sour-grapes aberration that occurs all too naturally in people afflicted with congenital incapability of one sort or another.

Nevertheless, I. came around to the view that there was something wrong with topology.

As I worked out a conceptual system suitable for my own purposes, I became more and more aware of ways in which accepted mathematical teachings seemed to me to be in violation of good sense. I soon had to ask myself whether I was afflicted with a wasting mental disease that was destroying little by little all my faith in mathematics and its underlying logic. Perhaps I had mathematical leprosy and could expect to be treated accordingly if my diseased thinking became known.

While still brooding over my predicament I had occasion to make a cursory survey of the literature dealing with frauds and hoaxes perpetrated in the name of science and scholarship. The amount of human effort and ingenuity that has been expended on such fabrications is so impressive as to invite comparison with the amount expended in scientific and scholarly good faith, often to no less worthless ends. It occurred to me in this connection that various theories and representations put forward by scientists and scholars had amounted to "unintentional hoaxes," if I may be excused for making use of such a self-contradictory expression. At once the topology of point sets leapt to mind as the perfect example of such an unintentional hoax.

Further reflection led me to speculate that a very large part of modern mathematics might consist of similar unintentional hoaxes. Indeed the whole of modern mathematics in its "pure" or nonempirical embodiment might be suspected of being an unintentional hoax of vast scope and extravagant elaboration. Clearly, any definitely intended mathematical hoax would be indistinguishable as such if it carefully followed the ground rules currently laid down for playing games in the mathematical field. Mathematicians learn to regard incongruity of meaning, not as something laughable, but as something inexcusable or irrelevant for their purposes, depending on whether the incongruity occurs inside or outside a train of thought that purports to be free from all such incongruities. Whether there is anything else peculiar about the sense of humor of mathematicians is a good question.

Once again I. had a point, and not just the one about mathematicians' sense of humor. People in general think that mathematics is worthy of study and support because it is useful. However, a large amount of the mathematical enterprise is what I. called "nonempirical" and is not useful in the sense that people think of as being useful. When pressed to justify studying things like packing spheres in 24-dimensional space, mathematicians will sometimes assert that what they are investigating may someday turn out to be useful and will cite the list of discoveries made for their own sakes that later turned out to be just the thing that the physicists or chemists needed. I. knew about this argument, but he was not impressed:

> Mathematics has justified disregard for the empirical applicability of its intellectual constructions by claiming in effect that it is engaged, from the scientific point of view, in stockpiling intellectual constructions that might conceivably serve as mathematical models of something empirically known or knowable. As a justification for the unminding multiplication without limit of mathematical excogitations very few of which can ever be empirically useful this pleading is palpably ridiculous. It is merely a particularization of the argument that all knowledge, or at least all generalization of knowledge, deserves respectful preservation because some of it might someday be useful. The preservation of all ungeneralized knowledge or any large part thereof is a patent impossibility. No less impossible, if not so patently, must be the preservation of all generalized knowledge when such knowledge is subject to galloping overproduction.
>
> The cardinal tenet of the mathematician's faith is irreconcilable with the cardinal tenet of science, according to which there is discernable self-consistent order in the universe. The weight of evidence supporting the scientific outlook on the universe is so overwhelming that the claim of mathematics to be exempt from empirical justification can only be regarded as akin to the claim of revealed religion. Both mathematics and religion are rooted in the world of experience but, having resorted in the face of seemingly inexplicable mysteries to claiming exemption from submission to the test of mundane experience, they are forced to rely on the proliferation of nephelococcygic subtleties for the defense of the positions in which they have placed themselves.

Should mathematics follow the line of subjects such as classics and literature, give up assertions about dollars-and-cents usefulness, and justify itself on the grounds of being a study worthy for its own sake? Perhaps we do not need to consider the question, since we can dismiss I. as being not worth listening to because he was a crank:

> The "establishment" ruling that empire [of "mathematical superintellectualism"] would in ordinary times not even become conscious of my expression of dissent. If it did somehow, its instinct would be to give what I have to say the silent treatment. In these extraordinary times, however, the lords of postulation will readily appreciate that my allegations may make them no longer immune from the attentions of revolutionary nihilism. A corrupted mathematics has polluted the thinking of science, which has been imposed on the masses as a superstition, giving us a falsified civilization, now putrescent. Our mathematics is the most distinctive feature of our civilization, and in it the focal point of infection is to be found.

SOLUTION TO A PUZZLE

Earlier there appeared the problem of explaining how S. T. was able to go from

$$X = 1\,2\,3\,4\,5\,6\,7\,8\,9$$

to

$$X^2 = 1\,4\,9\,7\,7\,9\,4\,1\,9.$$

The solution is that T. squared digit by digit and added the digits in each square. So, the first five digits in X^2 come from

$$1^2 = 1, \quad 2^2 = 4, \quad 3^2 = 9; \qquad 4^2 = 16, \quad 1 + 6 = 7; \qquad 5^2 = 25, \quad 2 + 5 = 7.$$

The last four are derived in the same way. Squaring is simple when you know how.

SPHERE, PHILOSOPHY OF THE

A nice idea, that the sphere should have a philosophy. The subtitle of

The Point's Synthesis

by I. V., a 77-page paperbound book, is

The Sphere's Philosophy.

The book, "profusely illustrated," was published by a vanity press in 1984 with a price of $50 on the back cover. Also on the back cover is a picture of its author from the shoulders up, a young man dressed in a trenchcoat with its collar turned up, staring with seriousness and intensity at a point just above the camera. The back cover also contains what would be the dust-jacket copy if the book were hardbound and had a dust jacket:

> *The Point's Synthesis* presents the sphere's philosophy, which was lost by mankind in the agitated times of the destruction of the Phoenician civilization by the rise of the Hellenistic civilization. It proves the fact that the Arab decimal system is the ancient secret code of the sphere's philosophy and that the numbers are not abstract marks but are the elements of the sphere's philosophy.
>
> Searching for the point's synthesis, [V.] reached a philosophical move that runs through the sphere, and he observed the living move of the circle, then decreased the circle until it reached the point's circumference. He was able to activate the point through the number's graphics, which hide the circle's life. By building himself the spheres and following the discovered move, the author had the opportunity to discover the existence of an unknown second zero in the science of mathematics—the antizero.

This book should have been well over 500 pages, but, unable to publish the manuscript in its entirety, the author reduced it to its present size. Using the many drawings and tables in hope that those interested would be able to "read" them, the author has communicated his discoveries to fourteen academies of science.

Only thirteen were listed, though: those of Sweden, India, Japan, Italy, Israel, Germany, France, Australia, Denmark, Greece, England, Egypt, and Spain. The National Academy of Science of the United States might feel slighted, but to make up for it V. sent a copy of his book to fifty universities, exactly one in each of the fifty states. He made some interesting choices. That Arizona State University was chosen for Arizona's copy rather than the University of Arizona should get neither school very excited, but the University of Texas at Austin has a right to feel a bit miffed that it was passed over for the University of Houston. Duke University prevailed over the University of North Carolina, perhaps understandable; but what did Georgia State University have that the University of Georgia, Emory University, or the Georgia Institute of Technology did not? Princeton got New Jersey's copy, but in Massachusetts Harvard had to take a back seat to the University of Massachusetts.

The book is not dull. It starts dramatically:

> They were Semitics and left the Arabian Peninsula circa 2500 B.C. In waves of migration, they took to the roads of Mesopotamia east of the Mediterranean Sea. Among them were the Arabs, the Accadians of ancient Babylonia, the Assyrians, the Armaceans, and a part of Ethiopians.

You could not stop reading there, could you? The chapter headed

THE CIRCLE IS THE POINT'S CIRCUMFERENCE

starts:

> Take a ruler in your hand. It has two ends, the beginning, at the place where the numbers begin and the second end at the place where the marking of numbers ends. Look at the ruler's beginning. There is the house of zero. You always put that zero at the beginning when you measure the size of anything.
>
> It's morning. You go to your place of employment. You leave your house. Then you walk all day. In the evening you come back to the same house which you left. You go on again for a day or so to another place and then you come back home. How are you counting all your steps? Do you begin by counting from your house? Yes, if you start counting your house as Zero Steps. No—if you count the first step only after you did it. Then you come back in your house.

The contents of the book do not live up to its promise. V. in a sense derived our usual numerals from circles, created a mysterious anti-zero called z with the properties that

$$0 + 0 = z, \quad 1 + 0 = z, \quad 2 + 0 = z, \ldots, \quad 9 + 0 = z,$$

and after a while lapsed into crankese:

[0 + 0 = z is] the PLACE where lives the hidden Second Zero and where unseen lives the Commutation of Zeroes. The development can't be seen in a theoretical way but only through following The Sphere's Philosophy signals and building The Sphere's of The Sum's Table of The Sphere's Philosophy. The Commutation of Zero takes place through the turning of The Sphere's Mobile Circle through all Commutations of The Numbers of The Arab Decimal Systems. The Sphere's Philosophy is the only way through which the Commutation of Zeroes can be seen and proved.

I know that you have been waiting for a description of The Sphere's Philosophy, but you will not get it. It is not because I am withholding it, but that in all of the book's 77 pages it is nowhere defined, either clearly or not. As far as I can tell, it has something to do with how our Hindu-Arabic numerals are drawn:

You, mathematicians, are not descendants from the real founders of mathematics but, all of you are the descendants of the Greek's error of taking the ball as a sphere. The origin of the decimal system does not come from fingers and toes but from the sphere's philosophy. The Numbers' Graphics is not the abstract art of someone but is the great work of the founders of the sphere's philosophy. Their graphics tell us about the relations which exist between the Sphere's Circles and the sets of numbers.

If you want to find out more, you will have to spend the $50.

STATISTICS, PARAMETER ESTIMATION IN

There are no famous unsolved problems of statistics and the subject is not taught to everyone in high school, so you might think that there would be no statistics cranks. Not so: no matter what the field, you can usually find a crank in it. Here is R. T., writing of himself in the third person:

> The statistician proposes a new theory of parameter estimation. It is expected to result in an amicable solution to the Neo-Bayesian vs. Orthodox controversy. It is in fact expected to replace the Orthodox theory, and to have widespread (if not profound) effects on mathematical statistics. Perhaps everyday practical statistics, including the use of Bayes' formula, will remain unchanged for the most part, but the theoretical outlook will be different. The new theory will tend to unify statistical theory and to raise it to a level of mathematical maturity on a par with probability theory itself. To a statistician who has struggled in the reliability-demonstration field for ten years, the new theory appears awesomely beautiful. He thinks he is the first to have seen it, and feels much as Balboa must have felt when he first saw the Pacific Ocean.

What is it? Let us see it! Probability theory is lovely, a coherent whole proceeding from a few axioms with theorems both pretty and deep, and all the more admirable for being applicable. Statistical theory, on the other hand, strikes many mathematicians as being a patchwork of this and that, with ad hoc solutions to isolated problems, no unity, and no beauty beyond that of a steel girder bridge: it does its job in a utilitarian fashion, but the rivets show. A beautiful theory of statistics would be welcome. However, T. did not let us see it because he was looking for support:

Perusal of a typical statistical journal indicates that many papers were supported by various sponsors. Yet, the contents of many would be very different if the new theory were known to the authors. The statistician believes this situation puts the development of the new theory as among the first matters for consideration in mathematical statistics today. He seeks a minimum of one year undisturbed time for the development of this theory. It is realized, of course, that a potential sponsor may not have faith in an unknown statistician. Therefore the statistician is willing to consider no risk sponsorship(s), in which the sponsor(s) need not be held financially responsible if the statistician should not be able to substantiate his claim. Details have not been worked out. At the moment the existence of any sponsor at all needs to be established. It is hoped that statisticians reading this paper will bring it to the attention of potential sponsors or employers, and even to suggest that there is good reason to cut red tape for this case; that there does exist an emergency. . . .

It may be mentioned in passing that industrial organizations (among others) desiring to extract all the relevant information from expensive tests and/or available data cannot afford to ignore this theory. Huge sums are spent even today in the reliability-demonstration field.

T. needed time:

The respective claims of the shortcomings of the two theories, as evidenced in the controversy, will be collected for the purpose of testing the new theory against these shortcomings. It is expected to have none of these shortcomings. Philosophical issues concerning the use of subjective knowledge will be investigated to the extent that they affect the theory. Clarification of the problems surrounding this area is expected.

About half the time may be spent in the writeup. This includes defining concepts, formulating and proving theorems, and solving famous problems according to this theory. Appropriate illustrative problems will be considered. Multiple parameter estimation is to be included. If time allows, the embryonic development of some new fields will be attempted.

As far as I know, he did not get the support that he wanted. In any event, there has been no change in the contents of the chapters on parameter estimation in statistics textbooks. T. might have had better luck if he had not used the passive voice quite so relentlessly.

What he had in mind we will probably never know. His paper ended with:

Please address all correspondence to [T.].

If questions do not relate to sponsorship or employment please send a self-addressed stamped envelope (the statistician has no secretary). This paper contains all the information about the theory the statistician is willing to part with at this time.

TAXONOMY, MATHEMATICAL

The letterhead read

FUNDAMENTAL CONCEPTS RESEARCH

D. S., proprietor,

Author of "Orthogonic Taxonomy of Fundamental Concepts"
Specializing in classification, based on
correlation to the kingdom category
with consistent significance of the digits

A lot of thought had gone into the *Orthogonic Taxonomy*, as a portion of page 104 shows:

    *333 boOo

TRANSCENDENTAL FUNCTION; mathematical analogy; function with an axis transverse to the axis of range and the axis of rank; concretion of the high correlation bases (or elements) of mathematics.

    *3330 boOoa

LOGARITHMIC FUNCTION; analogy of digital space;

    *33300 booOoaA Briggsian antilog$_{10}(N)_{10}$; digital hierarchy; N

    *33301 boOoar antilog$_e$; exponential function

    *33302 booOoae Briggsian log$_{10}(N)_{10}$; cyclic interpolation

    *33303 booOoao logarithm to the base e, (boOaoa); log$_e$

    *33304 boOoau evolvement of Nth root; (of cyclic radix)

    *3331 boOor

TRIGONOMETRIC FUNCTION; D514; analogy of angular space;

    *33310 boOora $\log_i^4(\text{angle})_4$; natural angular measure

*33311   boOorR   radian measure
*33312   boOore $\cos\theta$; sine; x/h_2
*33313   boOoro $\cosh\theta$; hyperbolic cosine
*33314   boOoru   hs; hypotenuse;
$(h_4^2 = -0^2 + x^2 + y^2 + z^2 - c^2t^2)$; hD
*3332   boOoe
PRODUCT FUNCTION; analogy of factor space;
*33320   boOoea
$z! = z(z-1)! = 1 \cdot 2 \cdot 3 .. z$; $\Gamma(x) = x\Gamma(x-1)$

It goes on in similar fashion. S., who called himself "rational fundamentalist, g.p." seemed to have bitten off more than he could chew: even though he had taxonomized *all* of mathematics, he did not know all there was to know about it:

> I have derived a method of writing very large numbers, and I am wondering whether this method gives larger numbers than any other method.
> I discovered this method in the routine process of deriving the categories of my periodic table of fundamental concepts.

His method was to write a repeated exponential, such as

$$4^{5^6},$$

as $4 \sqsupset 5 \sqsupset 6$, and then to introduce a new notation for the exponential analogue of factorial:

$$\boxed{6\,|},$$

which stands for the exponential tower

$$6^{5^{4^{3^{2^1}}}}.$$

Those numbers get big pretty fast ($\boxed{5\,|}$ already has 183230 digits), but anyone who presumed to know enough to be able to classify all knowledge about all of mathematics should not have

> to enquire of you whether or not there are other methods that give larger numbers than mine.

Of course there are. Any monster-number generator that S. devised I can beat, by putting "!" at the end of it. Or, for that matter, "+1."

TIME, WASTED

Can you stand to have your heart wrung? In 1956, A. M. circulated a four-page printed pamphlet summarizing his manuscript, *Dexsinal Gauges*.

> This thesis is in legible longhand and contains 207 pages. It is the fruit of 36 years (1918–1954) research. The author, [A. M.], A.B. [I.] University 1903, M.S. [W.] University 1936, desires to offer this thesis in total fulfillment for an honorary Ph.D. degree in Mathematics....
>
> This thesis:—
>
> 1. Explains the independent operations of dexsination and sindexation. It explains how to calculate fractional and integral gauges. This difficult problem was akin to that facing John Napier in his calculation of logarithms. Just as you multiply by adding logarithms and divide by subtracting logarithms so you dexinate by adding gauges and sindexate by substracting gauges.

"Substracting" is as printed. Other of the pamphlet's numbered points included:

> 3. It explains dexsinroots.
> 5. It explains the reciprocal law of density in Geometry as bearing on addition, multiplication and dexsination rectangles.
> 9. It shows a dexsinal relation between e and pi.
> 10. It explains the dexratio.
> 11. An equality of dexratios gives a poise. The poise has theorems similar to those about proportion. We find mean poisal of two numbers third poisal, fourth poisal, etc.

13. The thesis explains the laws of basal coincidence.
15. Other topics are Poisal isosceles triangle, the balance, other bases than 2, dexinal equations, commutative factors, heavy numbers, etc.

The pamphlet gives no hint of exactly what dexsinals were, but I am sure they are of no interest whatsoever to anyone except their creator. Thirty-six years of labor!

The author cannot attend the university. His wife is an invalid 15 years diabetes and at present slowly recuperating from a broken hip.

TOPOLOGY, APPLIED

There follows the complete contents of a letter to the mathematics department of a major eastern university, with its second sentence omitted for the moment.

> I believe that topology can be used as a mapping of chemical bonding and structure. . . . In general, it can be applied to the structure, interdependence, motion and other interrelationships of matter and energy. It has future uses in nuclear chemistry and physics, nuclear fusion, space travel and anything involving structure.
>
> Please acknowledge.

The reason for including a letter with such little content is its second sentence, which is:

> I have been told that I look like a topologist.

You might want to reflect on that for a moment, especially if you are a topologist.

TRISECTION OF THE ANGLE

This is the last in alphabetical order of the three unsolved problems of Greek geometry—the duplication of the cube, the squaring of the circle, and the trisection of the angle.

While the trisection cannot be done with straightedge and compass alone, it can be accomplished using a common tool, a protractor. It can also be done using a compass and a straightedge with two marks on it. Archimedes found this construction long ago (Figure 47): put a circle, with radius equal to the distance between the marks of the straightedge, around the angle to be trisected, and extend OB to the left. Then, put the straightedge down so it passes through A and one of the marks is on OB. Then slide the straightedge around until the other mark lies on the circle, at D. Angle ACB is then one-third of angle AOB. Why this is so is an exercise suitable for geometry students.

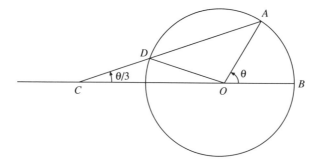

FIGURE 47

342

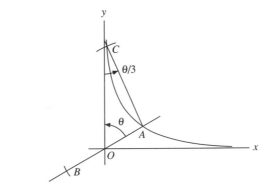

FIGURE 48

The ancient Greeks found other methods for trisecting angles, and modern methods are still being uncovered. Here is one, found in 1989. It may duplicate 2000-year-old work by the Greeks, but it was new to its discoverer (Figure 48). Take a hyperbola $xy = 1$ and put the angle to be trisected at the origin, as shown. Make $|OB| = |OA|$, and locate C by striking an arc with center A and radius $|AB|$. Extend AC to the y axis and you get the trisection. Why *this* is so is an exercise for students of analytic geometry.

The reason the trisection is impossible is that the construction is equivalent to solving a cubic equation with straightedge and compass. Constructions with straightedge and compass produce only square roots, and no matter how many square roots you take you can never get a cube root. Cube roots are needed to solve cubics. The difficulty with the proof that cube roots cannot be constructed with straightedge and compass alone is that it is beyond most trisectors and thus does not make any impression on them.

The subject of trisection was covered thoroughly in a 1987 book, so there is no need to go into it at great length here. The book's author expressed the hope that its appearance, exhibiting as it did the folly of attempting the trisection, would discourage or even eliminate the efforts of amateurs to trisect. Discouragement there may have been, but elimination seems to be impossible: I have received letters saying, in effect, "I have read a book on trisections. It was very interesting. Here is *my* exact trisection with straightedge and compass alone. What do you think of it?"

There are some things that did not appear in that book that might be of interest. Here are a few.

Albrecht Dürer (1471–1528) was northern Europe's first Renaissance man. While he is best known as an artist, he did a little of everything. He is known to mathematicians as the creator of "Melancholia," that engraving of an angel-winged creature staring into the middle distance, head wearily resting on one hand while the other hand loosely holds a compass, not seeing the rainbow in the distance, with sphere and starving dog at her feet, and a huge semi-dodecahedron gathering dust in the center of the picture. In the upper right-hand corner appears a magic square,

$$
\begin{array}{cccc}
16 & 3 & 2 & 13 \\
5 & 10 & 11 & 8 \\
9 & 6 & 7 & 12 \\
4 & 15 & 14 & 1,
\end{array}
$$

which exhibits in the middle of its bottom row the date the engraving was made, 1514. Off-hand brilliance like that is appealing to mathematicians, so appealing that the engraving keeps getting reprinted in histories of mathematics despite its plain message that too much learning, including mathematical learning, brings weariness of the flesh, depression, and scholars' melancholy.

Another example of Dürer's achievements, known hardly at all, is his trisection of the angle. In a pamphlet entitled

The
Printer's Manual

A Manual of Measurement of Lines, Areas,
and Solids by Means of Compass and Ruler
Assembled by Albrecht Dürer for the Use
of All Lovers of Art with Appropriate
Illustrations Arranged to be Printed in
the Year MDXXV

Dürer showed how to inscribe a square inside a circle, how to draw pentagons, hexagons, and nonagons, how to make tile patterns, and how to do other constructions with straightedge and compass. He made no distinction between those that can be done with perfect accuracy, like making pentagons, and those that cannot, like constructing nonagons.

His trisection of the angle—"Method for dividing an arc into three equal parts," he called it—is, of course, approximate, but it is one of the best approximations ever made. Also, it is not hard to do, as would be fitting in a book for practical artists (Figure 49). Take the angle to be trisected, BOA, or the arc to be trisected, BA, join A and B, and trisect that segment at C. Erect a perpendicular at C, meeting the arc at D. Draw an arc from D with radius $|AD|$ meeting AB at E. Trisect EC to determine F. Draw an arc from F with radius $|AF|$ to intersect the arc AB at T, the trisection point.

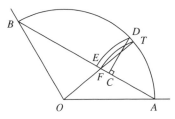

FIGURE 49

Look at how good that construction is:

θ	Dürer's $\theta/3$	Error
9°	3° 0′ 12.9″	12.9″
18°	5° 59′ 56.8″	1.2″
27°	8° 59′ 59.7″	.3″
36°	12° 0′ 0.1″	.1″
45°	14° 59′ 59.8″	.2″
54°	17° 59′ 59.7″	.3″
63°	20° 59′ 58.7″	1.3″
72°	23° 59′ 56.8″	3.2″
81°	26° 59′ 52.6″	7.4″

It is an order of magnitude better than the constructions usually found by trisectors. If you were aiming a missile from North Dakota to Moscow (I know that we no longer aim missiles at Moscow, but when examples are needed old habits die hard) and made an error of .2″, your missile would land approximately fifty feet from where you aimed it, quite close enough for practical purposes.

Dürer did not say where the construction came from. His translator said that Dürer's construction for a nonagon was "taken over from the traditional procedures of builders," so perhaps his trisection was also. On the other hand, it is hard to know how or why a tradition of nonagon-construction or trisections would have existed, and even in the sixteenth century the idea of the protractor was known. I think that it was devised by a scholar (while not suffering from melancholia) for its own sake, such being the attractive power of mathematics. The scholar could have been Dürer.

Here is how to handle trisectors, how *really* to handle trisectors. In 1988, D. L., of California, whose letterhead gave his specialties as

Systems Design
Technical Writing
Photo-Media Counsel
Data Resources
Applied Creativity

wrote to one of the bigger guns of American mathematics about his trisection:

> Am I only trying to sail against the usual gale of doctrinaire rebuff by making this admission and inquiry? It would be good to have some sustained dialogue beyond the conditioned reflex presumption of impossibility. *I have solved* ruler/compass trisection for the general angle, and I know the whole script about how it has been proven impossible to accomplish this because the problem is one of an irreducible cubic equation etc., etc. (I have read Courant and Robbins, repeatedly. Klein also. In fact, there is a copy of each work at chairside, for reading or reference, at my pleasure.)
>
> ... What is *most* desirable to me is finding a *publisher* for my *book* on ruler/compass trisection, now in preparation; and a financial sponsor or champion to enable completion and delivery of the exposition into distribution and validation.
>
> ... For me, writing an unprecedented book on mathematics is an activity that can only preclude or displace *other* productive pursuits competitively. But the sense of mission is compelling. Most assuredly, I am not a professing mathematician. If I were, the professional conditioning would have most likely left me anesthetized to the insights that led unerringly to the real solution. I'd have been trained to not look.
>
> This discovery ought to galvanize mathematicians world-wide, set off a brilliant era of new energy. Imagine what a celebration Pythagoras would have held if *he* had discovered this! It ought to be at least as widely heralded and splendidly rewarding. ...

And so on.

Here is the bigger gun's reply. I look on it with admiration.

Dear Mr. [L.]:

> If when you say "you know the whole script" you really knew it then you would not waste time on the problem. I really do know the whole script and I can assure you that [no one] is interested in what you have written.
>
> Now stop that stuff and get onto something useful. Think of all the real mathematics you could have learned during those hundreds or thousands of useless hours.
>
> As an aside, I wonder how anyone who cannot copy my name correctly can do careful mathematics.
>
> Reply to this letter if and only if it is to report that you have abandoned your fixation and embarked on something serious.

The danger in such forthrightness is that there is the small probability that the recipient of the letter may be moved to make his next communication a package that explodes on being opened. One must balance forthrightness with prudence. On the other hand, the last time I saw the bigger gun he seemed to be in fine health, probably all the finer for not undermining it by being polite to trisectors.

Most trisectors are old men—old because it takes a lot of time to trisect and because when one is young there are other things to do that are of more interest and importance. I think that trisectors are people of unsatisfied ambition, and the trisection is their last attempt at fulfilling it. Now and then, though, a trisector appears who is not old but in the prime of life. However, such trisectors usually have, for one reason or another, a good deal of spare time to devote to the trisection. Here, for example, is T. C., writing in 1990:

> I am a mental patient detained on a criminal warrant. Don't let that scare you because my illness was caused by a dormant form of diabetes only discovered when I did my crime a few years ago. However, the doctors are cautious because it still makes me act strange sometimes. Myself, I know I'm better and ready to be released, at least I am not a danger to myself or others. If I get recognition for some of the work I have done in mathematics and science it will help me get out or make my life better here by getting more privileges. I plan on taking a bookkeeping and computer course if I can get more privileges. I hope to prove I have the intelligence one needs for the course.

> I am 45 years of age, with a grade 12 formal education and very knowledgeable on many things because of my own studying. I would have got a better education but my illness made it difficult for me to function socially. However, it did make me intelligent in a certain way I'm not going into here. I spent my time studying the base of the universe, basically the Unified Field. I think I solved it with THE THREE PROBLEMS OF ANTIQUITY as the mathematical proof. It can all be understood by anyone with a high school education so don't worry of getting into some quack theory that is unintelligible. Concerning this, trisection of an angle is considered quackery. Part of the reason I have included 3 letters from [X. Y.], the head of the math department of the University of [Z.], in order to show you that it is of interest (my trisection theories) and worthwhile instead of just being chucked in the garbage which many consider the only place for trisection papers.

C.'s trisection is one that has been discovered before (see Figure 50): strike off arcs at distances of 2, 3, and 4 from the vertex of the angle, bisect the angle twice,

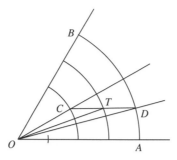

FIGURE 50

and connect C and D to get the trisection point T. It is not very accurate, with an error of $10'$ when applied to a $60°$ angle and $32'$ for a right angle.

Professor Y. had told C. that perhaps his proof that his construction trisected angles might be of interest as an example of fallacious geometrical reasoning. This C. interpreted as saying that his work was worthwhile. Poor C.!

VAN DER POL'S EQUATION

The subject of differential equations is mature; it is not a hot area of mathematics. The National Science Foundation sponsors no workshops for teachers to learn about new developments in the subject, publishers do not announce the founding of new journals in the subject, and departments of mathematics do not compete for young specialists in differential equations to round out their faculties. Useful and dependable differential equations are but stodgy and a bit dull.

It is not a subject that you would think would attract cranks. People have to know quite a lot of mathematics to have even heard of differential equations, and the more mathematics people know the less likely they are to be mathematical cranks. Also, there are no controversies in differential equations, no crises at their foundations, no school of non-standard differential equations: there is nothing to disturb the calm but dusty surface of the subject and nothing to attract cranks.

However, L. J. wrote in the foreword to his book, *The Stupid, Ridiculous Oversight*,

> The solution of differential equations by power series has never been advanced beyond its original primitive state because of a stupid, ridiculous, oversight.

It is not a book that I own since the author was asking $20 a copy, too much for me to pay. The two-page teaser that he distributed went on:

> Although it has been said that such solutions are to be found in the nature of the constants of the solution, it has never been understood how to "avoid" involving the independent variable in the process of obtaining these constants. This may be seen in past solutions of Legendre's and

Bessel's differential equations. How to "avoid" this is what this book is all about.

That was sufficiently obscure to make it almost certain that the book was not worth $20. J. went on:

> The absence of the independent variable in the process of solving differential equations in power series is of enormous advantage and leads to new linear solutions previously unknown. It also makes possible the exact solution of the well-known non-linear differential equations which have previously been solved only by approximation.

No independent variable in the solution process? That is, solve a differential equation like

$$\left(\frac{dy}{dx}\right)^3 + y^2 = 2$$

without using any x's? What could J. have meant?

He continued:

> Furthermore, the matter of rate of convergence of the power series solution of a differential equation is now (for both linear and non-linear differential equations) a matter of choice for the first time.

He concluded with:

> All of the advantages listed above are based upon the simple mathematical relation shown upon the outside of the front cover of this book. Why this simple relation and its results have not been known a long time ago is both a mystery and a scandal which deserves some thought. The consequences of this ridiculous oversight are obviously not small.
>
> This book contains:
> 1. The various power series used to solve exactly both linear and non-linear differential equations.
> 2. The exact solution of various linear and non-linear differential equations by these series.
> 3. The exact solution of both linear and non-linear differential equations by a choice of power series of as great a rate of convergence as one chooses.
> 4. How to solve for (a) in a power series [of $(x + a)$] solution of a linear differential equation.

5. If Yukawa's solution of his wave equation is assumed to be the muon, then two more solutions of non-linear differential equations derived from Yukawa's wave equation gives the other mesons.
6. A power series solution of Schroedinger's wave equation to give the energy levels of the hydrogen.

This book is easily understood by anyone with a first course in differential equations.

You may be think that it was shortsighted not to shell out $20 for something that promised so much. However, consider the following before you go seeking out J. to press cash on him. I have a sheet that J. produced to correct an error on page 156 of his book. It is headed

Exact solution of the Van der Pol
differential equation with forcing function

The van der Pol equation is usually written in the form

$$\frac{d^2y}{dx^2} + \mu(x^2 - 1)\frac{dy}{dx} + y = 0.$$

It has many picturesque solutions (Figure 51 shows one for $\mu = 10$) though it cannot be solved exactly in closed form. Nevertheless, J. said that for his version of the equation,

$$\ddot{y} + A\dot{y} + By^2\dot{y} + Cy = -Di\exp(i\omega t),$$

the solution is

$$y = [\cos(\omega t)]\sum_{n=0}^{n=\infty}(n/n!)(\omega t)^n.$$

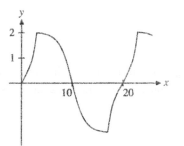

FIGURE 51

To apply that, we need to know what $\underline{n}$ is, and J. told us that

$$\underline{0} = 1, \qquad \underline{1} = -(A+B)\frac{\omega^{-1}}{2} - D\frac{\omega^{-2}}{2},$$

$$\underline{2} = 1 + [(A+B)^2 - C]\omega^{-2} + D(A+B)\frac{\omega^{-3}}{2},$$

$$\underline{3} = 3\underline{1} - [2(A+B) + 2B\underline{11} - A - 3B]\omega^{-1} + (D - C\underline{1})\omega^{-2},$$

and

$$\underline{4} = [A(3\underline{1} - \underline{3})\omega - B(\underline{3} - 27\underline{1} + 6\underline{12}$$
$$+ 2\underline{111})\omega - C(\underline{2} - 1)]\omega^{-2} + 6\underline{2} - 1, \ldots.$$

The "..." does not help much, but never mind.

Mathematicians are trained to examine extreme cases, and nothing could be more extreme than putting

$$A = B = C = D = 0$$

in the original equation, leaving

$$\ddot{y} = 0,$$

an equation whose solution is

$$y = c_1 t + c_2.$$

But when we substitute those values to find n, we get

$$\underline{0} = 1, \quad \underline{1} = 0, \quad \underline{2} = 1, \quad \underline{3} = 0, \quad \underline{4} = 5,$$

giving

$$y = (\cos \omega t)\left(1 + \frac{1}{2!}(\omega t)^2 + \frac{5}{4!}(\omega t)^4 + \cdots\right),$$

which, no matter what appears in the rest of the series, does not agree very well with the other solution. J. must have committed an oversight, perhaps stupid and ridiculous.

NOTES

In the preceding text, few of the writers were identified by name. The cranks and eccentrics do not need to be named since their function is to serve as examples. What is important is what they did, and how and why they did it, but not who they were. In the same way, who the patient mathematicians were who corresponded with them is not important. In fact, some mathematicians might not want it known that they had spent their time in that way.

However, there is a possibility that someone may want, someday, to look into this subject that I have found so interesting, and for that reason this section will give some references. Of course, not every person referred to by initials in the text will be identified. Those writers who confined themselves to private correspondence—and there were many—must have their privacy respected. Proper procedure is also clear for those who produced a book, found a publisher for it, and put it on sale: they may be treated in the same way as any other author who places a book before the public.

But not all writers fall into such clear-cut categories. What is one to do with the author who mentions prominently that his work has been copyrighted, even to the extent of putting a little circle with a c inside it on each page? Some cranks fear that their work will be stolen and, even though it is circulated to only a few correspondents in hand-written form, think that it needs to be protected with that magic symbol. What I have decided to do is to consider mentioning an author's name if his work is both copyrighted and printed. The two together tend to indicate that he wished for his work to circulate among the public, and if so he cannot complain about public comment on it. Nevertheless, I decided that some printed and copyrighted work really belongs in the class of private communication, so it does not appear in this section. If an author's printed and copyrighted work included a price, then I had no hesitation in identifying him since anything for sale is by definition not private.

Similarly, cranks who succeed in getting a story about their work in the newspapers or in getting an article printed in a periodical *want* to be known and also cannot complain about being mentioned by name.

In addition, some books and articles referred to in the text without references are identified more fully here.

American Revolution, The Role of 57 in the

The tiny pamphlet, *History Computed*, was offered for sale at $2 a copy in 1983 by Arthur Finnessey, 1093 Country Lane, Atlanta, Georgia 30324.

Applied Mathematics

The author of *The Fundamental Equation of the Construction of the World*, published in 1939, was Lawrence Breuhl, of 509 Fifth Avenue, New York, New York. That pamphlet advertised the author's forthcoming 406-page book, *The Mathematics of Unlimited Prosperity*, but I do not know if it ever appeared. The deviser of Wilkinson's complete mathematics method, created around 1965–1970, was Austin Wilkinson of Seeley, California. He had a small display advertisement of his method in the classified section of his local telephone directory under four different headings: "Mathematicians," "Books," "Inventions," and "Schools."

Base of the Number System, The Best

F. A. stands for F. Emerson Andrews and G. T. for George Terry. Professor Gene Zirkel, an officer of the Dozenal Society of America, can be written to at the Mathematics Department, Nassau Community College, Garden City, Long Island, New York 11795. The author of "Count by eights, not by tens" was E. M. Tingley, and his article appeared in *School Science and Mathematics* in the April, 1934 issue, pages 395–401. W. F. Kemble, of Lower Bank, New Jersey, advocated his mixed base-24 and base-60 system in *Letter Systems in Business and Technology*, printed in 1942.

Cantor's Diagonal Process

The author of "A correction in set theory," in which Cantor was refuted, was William Dilworth, whose article appeared in the *Transactions of the Wisconsin Academy of Sciences, Arts and Letters* in 1974.

Congressional Record, Mathematics in the

The *Congressional Record* of June 5, 1940 contained the discoveries about π and e by Miff M. Butler, inserted by Representative Frank O. Horton of Wyoming. On June 3, 1960, the *Record* again contained mathematics, on pages A4733–4734, where Representative Daniel K. Inouye inserted mention of Maurice Kidjel and Kenneth W. K. Young, authors of *Challenging and Solving the "3 Impossibles,"* a 42-page pamphlet published by the Kidjel Ratio Division of the Hawaii Art Publishing Company in 1961 and distributed by Kidjel-Young and Associates, 1012 Piikoi Street, Honolulu, Hawaii. Representative Inouye's remarks concluded:

> The Kidjel ratio system textbook, the Kidjel ratio calipers, and the Kidjel solutions to the "three impossible problems" in Greek geometry are on display at my office for the inspection of my colleagues and members of their staffs, should they be interested.

The Kidjel quadrature of the circle would be correct if $\pi = 2 + \frac{2}{3}\sqrt{3}$, but that equation is false.

Constant Society, The

The advertisement for *Constant Processes* appeared on the back cover of the April, 1978 issue of *American Mathematical Monthly* (volume 85, number 5).

Crank, Case Study of a

All the Primitive Pythagorean Triples was published by the B. G. Michael Syndicate, 525 Page Avenue, NE, Atlanta, Georgia 30307 in 1976, was on sale at a price of $4.50, and was written by George M. Bright. The falsity of the conjecture $L_p \equiv 1 \pmod{p}$ if and only if p is prime, where L_n denotes the nth Lucas number, is mentioned in "Some congruences of the Fibonacci numbers modulo a prime p" by V. E. Hoggatt, Jr. and Marjorie Bicknell in *Mathematics Magazine*, **47**(4) (1974) 210–214.

Crank, The Making of a

The Four-Colour Conjecture, a 16-page pamphlet by William Shipman, was published in 1982 by Almonte Shipman Enterprises, P.O. Box 176, Station "A," Scarborough, Ontario M1K 5B9. There was a supplement in 1983.

Deduction, The Joy of

The International Society of Unified Science, supporters of the idea that all is motion, has its headquarters at 1195 S. Windsor Street, Salt Lake City, Utah 84105.

Duplication of the Cube

The duplication of the cube achieved by doubling the area of its faces can be found in *Solutions of the Three Historical Problems by Compass and Straightedge*, by Delvin J. Johnson, 27060 Cedar Road, Beachwood, Ohio 44122. The accurate duplication achieved by using a straightedge with two marks on it appears in *Duplicating the Cube* by Carleton C. Taylor, Jr., published by the Albuquerque Printing Co., Albuquerque, New Mexico, 1981.

Fermat's Last Theorem

Two excellent general references for FLT are *Fermat's Last Theorem* by Harold M. Edwards (Springer-Verlag, New York, 1977) and *13 Lectures on Fermat's Last Theorem* by Paulo Ribenboim (Springer-Verlag, New York, 1979). Shafi Ahmed's pamphlet *Proof of Fermat's Last Theorem* has had several editions. The first was published in 1972 by Lily Ahmed, 172 Western Avenue, London W.3. There was another in 1982 published by the author at 100 Western Avenue, London W3 7TX, and a revision appeared in 1988. His book, *The Absolute Theory of the Universe*, was published in 1980 by New Horizon, Horizon House, 5 Victoria Drive, Bognor Regis, West Sussex. Arnold Arnold's proof of FLT was written up in the *Guardian* of January 12, 1984, with the semi-retraction appearing on January 26. There were other mentions in the *Observer*, January 15, and the *New Statesman* dated January 20. J. C. Edwards, the Secretary of the British American Scientific Research Association and Editor of its *Journal*, can be reached at 49 Marsh Crescent, Regina, Saskatchewan S4S 5R3. *Convergence Surds and Fermat's Last Theorem* was written by Max M. Munk and published in 1977 by Vantage Press, 516 West 34th Street, New York, New York 10001. Correa Moylan Walsh's book, *An Attempted Proof of Fermat's Last Theorem by a New Method*, was published by G. E. Stechert and Company, New York, in 1932.

Fifth Postulate, Euclid's

I had never heard of Georg S. Klügel before reading about him on pages 867–869 of Morris Kline's *Mathematical Thought from Ancient to Modern Times* (Oxford University Press, New York, 1972). J. J. Callahan's 310-page book, *Euclid or Einstein*, was published in New York by The Devin-Adair Company in 1931. His pamphlet, *The Foucault Pendulum and the Newtonian Theory*, was published in 1975 by James D. Callahan, Box 43, Constance, Kentucky 41009. Matthew Ryan's pamphlet, *Euclid's "World-Renowned Parallel Postulate,"* was produced in 1905 by the Henry E. Wilkens Printing Co., 719–721 13th Street, N.W., Washington, D.C.

Four-Color Theorem, The

The Four Color Problem, by Robert L. Carroll, a 1977 publication of the Carroll Research Institute, was published by the J. R. Rowell Printing Co., 3839 Rivers Avenue, Charleston, South Carolina 29406. By the same author, appearing in the same year with the same publisher, is *The Four Solutions to Fermat's Last Theorem*. His *Eternity Equation* could be obtained from the same source.

In this section and elsewhere, reference is made to *A Budget of Paradoxes*, by the English mathematician Augustus De Morgan (1806–1871). This admirable work, originally published in 1872 and reprinted by Dover Publications in 1954, is now available from Ayer Co. Publishers, P.O. Box 958, Salem, New Hampshire 03079. It contains a collection of De Morgan's writing on a wide variety of subjects. By "paradox" De Morgan meant "something counter to general opinion," so his book contains many examples of cranks and eccentrics, not all in mathematics. It contains much else that is informative and delightful.

Goldbach Conjecture, The

The proof of Goldbach's conjecture consisting only of a sequence of equations is taken from a curious book, *Number Theory Seven*, by K. Savithri, published in 1986 by Sciencus Publications, Secunderabad, India.

Greed

Newton's Laws are Full of Flaws, by Al Snyder, was published in 1973 by the Snyder Institute of Research, 508 N. Pacific Coast Hwy., Redondo Beach, California 90277.

Legislating Pi

David Singmaster's marvelous article "The legal values of π" appeared in the *Mathematical Intelligencer*, **7** (1985) 69–72. It depended partly on "Indiana's squared circle" by Arthur E. Hallerberg, *Mathematics Magazine*, **50** (1977) 136–140.

Magic Squares

W. S. Andrew's *Magic Squares and Cubes* has been kept in print by Dover Publications. W. L. Schaaf's "Early books on magic squares" appeared in the *Journal of Recreational Mathematics*, **16** (1983–1984) 1–6. The square with magic constant 666 appeared in "666: The number of the beast," by A. W. Johnson, Jr., in the *Journal of Recreational Mathematics*, **16** (1983–1984) 247.

Megalomania

Dr. George N. Kayatta is a tireless self-promoter, always seeking access to the media. His accomplishments have been mentioned in the pages of the *New Lebanese American Journal* for 1/12/87, 3/9/87, 12/14/87, 1/11/88, 2/8/88, 3/14/88, 4/11/88, 8/12/89, and no doubt other dates as well. The story about him in the New York *Daily News* appeared on December 20, 1987, on page 18, and the item in the *Wall Street Journal* appeared in the issue of October 5, 1987, on page 25. Issues of *The New Renaissance* are published by New Renaissance Productions, Inc., 34 West 28th Street, New York, New York 10001.

Nines, Casting Out

The subtitle of *Instant Mathematics*, by Francis Soh—no publisher but "printed by Meko," Subang Jaya, Selangor, Malaysia, 1989—is *For Office & Commercial Applications. Specially for Teachers, Managers, Business Entrepreneurs, Supervisors, Cashiers, General Clerks, Etc.*

Notation, Nonstandard

The article referred to is "Tabular integration by parts" by David Horowitz, *College Mathematics Journal*, **21**(4) (1990) 307–311.

Number Theory, The Lure of

The issue of *Hypermodern* in which Allan B. Calhamer's conjectures appeared is the twelfth, dated November 17, 1970, and published at 501 N. Stone Avenue, La Grange Park, Illinois 60525.

Phi

The Divine Proportion by H. E. Huntley (Dover Publications, New York, 1970) is an entire book written about ϕ.

Primes, Twin, Existence of Infinitely Many

The professor who thought he had proved the twin primes conjecture was Charles N. Moore of the University of Cincinnati.

Pythagoreans, Neo-

Raymond Schulz's diagram appeared in the February 1968 issue of the *Mathematics Teacher* on page 176. His advertisement for the Pythagorean Society was in a later issue, on page 706, but I do not know the year.

Quadrature of the Circle

The amazing coincidences involving the digits of π were found by Monte Zerger and reported in "The magic of π," *Journal of Recreational Mathematics*, **12** (1979–1980) 21–23. *The Measure of the Circle*, "perfected in 1845," was published for the author, John Davis, in Providence, Rhode Island in 1854. The author of *Mathematical Commensuration*, whose full title is *A Prefatory Essay to the New Science: Mathematical Commensuration, Preceded by a Brief Retrospective View of Research in the Domain of Geometry*, was Chas. De Medici, "D.Ph.," and his 31-page booklet was published in Chicago by A. M. Flanagan in 1883. Sylvester C. Gould's bibliography *What is the Value of* π was published in Manchester, New Hampshire in 1888. *Quadrature of the Circle* by J. A. Parker is a 300-page book published by John Wiley & Sons, New York, 1874. The title of the book on locomotives by Zerah Colburn, also published by Wiley, was included because I was surprised to see the name of Colburn, one of the famous calculating prodigies of the past, in another context. Edward D. O. Moore was the author of *O. Moore's Geometrical Science*, about whose publication I have no more information than an indication that it was "Entered according to Act of Congress, in the year 1890, by Edward D. O. Moore, in the office of the Librarian of Congress." The subtitle of the pamphlet is *The Solution of this Great Problem Which has Baffled the Greatest Philosophers and the Brightest Minds of Ancient and Modern Times, from Pythagoras, Five Hundred Years Before Christ, to the Present Day, Has Now Been Solved by an Humble American Citizen, of the City of Brooklyn*. The article on Richard L. Emen, the

Kansas City circle-squarer, appeared in the Kansas City *Times* of July 12, 1983, complete with a picture of him, with his eyes bulging alarmingly. Lawrence Earl Cavender issued his *Unique Mathematical Geometrical Findings* in several forms from 1967 to 1973, but π was always $9/5 + \sqrt{9/5}$. David Bean's *A Collation of Geometrical Principles and Propositions and a Combination of Geometrical Figures by which the Ratio of the Diameter of the Circle to the Circumference is Found to be as 1 is to 3.25* is a pamphlet that was copyrighted in 1881, but I do not know by whom it was published. Albert Cottrell's *Cottrell's System and Process of Squaring the Circle* was published in Providence, Rhode Island in 1888 by Frank N. Shaw, Printer and Publisher. The circle-squarer who wanted to change the number of minutes in the hour was Bendicht Jacob of Stony Hill, Gasconade County, Missouri, who published his *Square of the Circle* in 1889. π—*A New Value* was first published in 1979, and in revised and extended form in 1981, by T. S. Jarnecki of the Civil Engineering Department of the Polytechnic of Central London. The 1931 book on the quadrature that is found in many college libraries was written by Carl Theodore Heisel, supporting the value of π found by Carl Theodore Faber. Dan W. Gaddy's 1988 pamphlet, *On the Exact Measurement and Quadrature of the Circle*, was published by Dorsett Printing Company, Rockingham, North Carolina, and was advertised for sale in the January/February 1990 issue of the *Mensa Bulletin* as being available from the author at Route 4, Box 553, Bennettsville, South Carolina 29512. Mr. Gaddy was written up in the Greensboro *News and Record* of March 20, 1989, in the Winston-Salem *Journal* of March 25, 1989, and in the Durham *Herald* of March 25, 1989. He also made an appearance in a Wake Forest College alumni publication, but I do not know exactly when.

Signs, The Rule of

A Challenging of Traditional Mathematics and Special Relativity by Dallas Irvine was published by Warren H. Green, Inc., 8356 Olive Boulevard, St. Louis, Missouri 63132 in 1981. In 1985 the same publisher issued a 15-page pamphlet by Mr. Irvine, *Science in Question*, "to replace pages 175–352" of the book.

Sphere, Philosophy of the

Ion Vulcanescu is the author of *The Point's Synthesis (The Sphere's Philosophy)*, published in 1984 by Exposition Press, 325 Rabro Drive, Smithtown, New York 11787.

Trisection of the Angle

The book by Albrecht Dürer, translated and with commentary by Walter L. Strauss, was published in 1977 by Abaris Books, 24 West 40th Street, New York, New York 10018. A proof that the angle cannot be trisected, "Elementary proof that some angles cannot be trisected by ruler and compass" by the eminent philosopher W. V. Quine, can be found in *Mathematics Magazine*, **63**(2) (1990) 95–105. Trisections and trisectors are dealt with in *A Budget of Trisections* by Underwood Dudley (Springer-Verlag, New York, 1987). One of the few books published by a trisector is *Trisection of the 120 Degree Angle* by George W. Kelly (Vantage Press, New York 1973).

Van der Pol's Equation

I have lost the coupon to fill out to send for Lloyd C. Jones's 274-page book *The Stupid, Ridiculous Oversight*.

INDEX

9436